Frank Mathias Hilker

Spatiotemporal patterns in models of biological invasion and epidemic spread

This publication contains the doctoral thesis of Frank M. Hilker, entitled *Spatiotemporal patterns in models of biological invasion and epidemic spread* and accepted by the Department of Mathematics & Computer Science of the University of Osnabrück in July 2005.

Die vorliegende Publikation enthält die vom Fachbereich Mathematik/Informatik der Universität Osnabrück im Juli 2005 angenommene Dissertation von Frank M. Hilker mit dem Titel *Spatiotemporal patterns in models of biological invasion and epidemic spread*.

Bibliografische Information Der Deutschen Bibliothek

Die Deutsche Bibliothek verzeichnet diese Publikation in der Deutschen Nationalbibliografie; detaillierte bibliografische Daten sind im Internet über http://dnb.ddb.de abrufbar.

ISBN 3-8325-1050-8

Logos Verlag Berlin
Comeniushof, Gubener Str. 47,
10243 Berlin
Tel.: +49 030 42 85 10 90
Fax: +49 030 42 85 10 92
INTERNET: http://www.logos-verlag.de

Abstract

Biological invasions are a severe ecological problem threatening biodiversity and causing substantial economic damages. Mathematical models of spatiotemporal spread have proven to be powerful tools in identifying the underlying mechanisms, thus contributing to the understanding of the factors that determine invasion processes and to the assessment of possible control methods. In this thesis, classical models are extended to combine spatial spread, population growth, disease transmission and community interactions. Applications are exemplarily found in the circulation of the Feline Immunodeficiency Virus (FIV) – an HIV-similar lentivirus that induces AIDS in cat populations – and in viral infections in phytoplankton that forms the basis for all food chains and webs in the sea. The joint interplay of epidemics, predation and environmental stochasticity in invasion models is shown to generate rich and novel patterns of spatiotemporal spread such as the blocking and reversal of invasion fronts or the spatial 'trapping' of infection as well as its noise-induced escape. The results of this thesis can explain real-world phenomena and have important implications for understanding and controlling invasion processes in ecosystems and epidemiology.

Zusammenfassung

Biologische Invasionen stellen ein großes ökologisches Problem dar. Sie bedrohen massiv die Biodiversität und verursachen erheblichen ökonomischen Schaden. Mathematische Modelle haben sich sehr bewährt, die zugrundeliegenden Mechanismen zu identifizieren. Auf diese Weise tragen sie dazu bei, sowohl die wesentlichen Faktoren zu verstehen, die den Verlauf einer Invasion bestimmen, als auch mögliche Kontrollmethoden zu bewerten. In dieser Arbeit werden klassische Modelle dahingehend erweitert, räumliche Ausbreitung, Wachstum, Übertragung von Infektionskrankheiten und Interaktionen mit anderen Spezies zu vereinen. Exemplarische Anwendungen finden sich in der Ausbreitung des felinen Immunschwächevirus (FIV) – einem HIV-ähnlichen Lentivirus, das zu AIDS bei Katzen führt. Ein weiteres Beispiel greift Virusinfektionen in Phytoplankton auf, das die Grundlage aller Nahrungsketten und -netze im Meer bildet. Es wird gezeigt, dass das gemeinsame Zusammenspiel von epidemischen Prozessen, Räuber-Beute-Interaktion und Umweltstochastizität in Invasionsmodellen reichhaltige und neue raumzeitliche Ausbreitungsmuster erzeugt, wie etwa das Aufhalten und Umkehren von Invasionsfronten oder die räumliche Eindämmung von Infektionen und ihre rausch-induzierte Ausbreitung. Die Ergebnisse dieser Arbeit liefern Erklärungen für in der Natur beobachtete Phänomene und haben weitreichende Konsequenzen für das allgemeine Verständnis und die Kontrolle von ökosystemaren und epidemiologischen Invasions- und Ausbreitungsprozessen.

Dedicated to the memory of my father

Published parts of the thesis

Parts of this thesis have been published or have been accepted for publication in international peer-reviewed scientific journals. They include:

Hilker, F. M., Langlais, M., Petrovskii, S. V., Malchow, H. A diffusive SI model with Allee effect and application to FIV. *Mathematical Biosciences*, accepted for publication.

Hilker, F. M., Lewis, M. A., Seno, H., Langlais, M., Malchow, H. (2005). Pathogens can slow down or reverse invasion fronts of their hosts. *Biological Invasions* 7(5), 817-832.

Malchow, H., Hilker, F. M., Petrovskii, S. V., Brauer, K. (2004). Oscillations and waves in a virally infected plankton system: Part I: The lysogenic stage. *Ecological Complexity* 1(3), 211-223.

Malchow, H., Hilker, F. M., Sarkar, R. R., Brauer, K. (2005). Spatiotemporal patterns in an excitable plankton system with lysogenic viral infection. *Mathematical and Computer Modelling* (in press).

Closely related publications include:

Malchow, H., Hilker, F. M., Petrovskii, S. V. (2004). Noise and productivity dependence of spatiotemporal pattern formation in a prey-predator system. *Discrete and Continuous Dynamical Systems B* 4(3), 705-711.

Malchow, H., Petrovskii, S. V., Hilker, F. M. (2003). Models of spatiotemporal pattern formation in plankton dynamics. *Nova Acta Leopoldina NF* 88(332), 325-340.

Petrovskii, S. V., Malchow, H., Hilker, F. M., Venturino, E. (2005). Patterns of patchy spread in deterministic and stochastic models of biological invasion and biological control. *Biological Invasions* 7(5), 771-793.

Note: The author has also published papers in other areas of theoretical population dynamics that are not directly linked to the content of this thesis.

Contents

II   Infections in prey-predator systems   69

4   Oscillations and waves in a virally infected plankton system Part I: The lysogenic stage   71

5   Spatiotemporal patterns in an excitable plankton system with lysogenic viral infection   89

6   Strange periodic attractors in a prey-predator system with infected prey   107

7   Oscillations and waves in a virally infected plankton system Part II: Transition from lysogeny to lysis   119

Chapter 1

Introduction

1.1 General introduction

Biological invasions are rapidly increasing worldwide. They are a major cause of species extinctions and economically expensive (Clavero and García-Berthou, 2005; Simberloff et al., 2005). Alien invaders are responsible for the loss of biodiversity by outcompeting indigeneous species and catastrophically damaging natural ecosystems (Lassuy, 1995; Sala et al., 2000). Infectious diseases threaten human health and are estimated to cause more deaths than wars and starvation. Pathogens and parasites threaten in many cases the persistence of animal and plant populations, which moreover may act as disease-vectors to humans. In US agriculture, damages caused by pests in crops and implemented control measures are estimated to cost US-$ 137 billion per year (Pimentel et al., 2000). The spread of genetically modified organisms and bio-engineered species as well as the potential misuse of alien invaders as weapons in bioterrorism have to be carefully assessed. The prospect of bacteria developing resistance against all available anti-biotics is highly terrifying.

In short, biological invasions are one of the most severe ecological problems. Unlike air and water pollution, biological invasions have an irreversible character for the simple fact that alien organisms reproduce. As an inevitable side effect of globalisation, they will become more and more relevant. The global trade and the accidental transport of non-indigenous species in ships and planes increase dramatically. Many human diseases have their origin in close contact between people and animals.

Consequently, there is a pressing need to better understand invasion processes (Vander Zanden, 2005). There is some ongoing effort from a more empirical point of view to assess the invasibility of ecosystems as well as which species may become invasive and which risk they pose (e.g. Ehrlich, 1986; Crawley, 1987; Mack et al., 2000; Heger and Trepl, 2003; Andersen et al., 2004). Nonetheless, tools are urgently needed to forecast future invader spread (Clark et al., 2001a) and to evaluate the worthiness of prevention effort (Leung et al., 2002). In particular, it has also to be taken into account that biological control is a difficult task and even may emerge as a risky affair (Simberloff and Stiling, 1996a,b; Courchamp et al., 1999b; Collier and Van Steenwyk, 2004; Hajek, 2004; Simberloff et al., 2005). In order to predict the course of invasions and implement striking control measures, however, it is necessary to understand the mechanisms and dynamics underlying biological invasions.

Mathematical modelling is an effective and essential tool to quantify the spatiotemporal spread and to reveal the fundamental "laws" that determine the fate of invasions. All the underlying biological factors act on different spatial and temporal scales. This heterogeneity induces mathematical difficulties, both from analytical and numerical point of view. Though much progress has been made in the last few years in identifying spread patterns, cf. mini-review 1.3 below, many aspects of considerable interest remain obscure.

Reaction-diffusion models have been successfully applied to explain the constant range expansion of invasive species (Fisher, 1937; Kolmogorov et al., 1937; Skellam, 1951). Within this modelling framework, the reaction terms describe the local population growth and the diffusion term the spatial spread. However, the interplay with other aspects from, e.g., ecology and epidemiology has not been properly and fully addressed yet. Hence, the challenge is to combine modelling approaches from different fields and to unify them. This allows to investigate the impact of additional factors on the invasion process. For instance: How do pathogens affect the fate of invasive species and can they stop the invasion of their host population? How does invasion take place when the invader is subject to a predator-prey relationship? What are the consequences, when the prey and/or the predator species are infected with a disease? This thesis sheds light on the following aspects as well as their joint interplay:

- spatial spread,

- more complicated population growth exhibiting a strong Allee effect, i.e., negative growth rate at small densities,

- epidemiology,

- trophic interactions of prey-predator type and

- environmental stochasticity.

In particular, the role of infectious diseases will be investigated. On the one hand, the spatial spread of epidemics and endemics is interesting in itself. On the other hand, the influence of introduced diseases on invading host populations is largely unknown yet. This thesis demonstrates that epidemiological aspects as well as the other above mentioned factors can significantly modify the spatiotemporal spread patterns of species. Invaders do not necessarily propagate with an average constant speed – they can be stopped, trapped, eradicated or slowed-down. Moreover, their spread can be patchy or occur as spiral waves.

Biological invasions can thus take place in a much more complicated manner than previously expected. The main contribution of this thesis is probably to highlight that the incorporation of additional, realistic factors enriches the patterns of spatiotemporal spread in a complex variety. This has, in its immediate turn, severe consequences for the design of monitoring and management strategies. The ultimate aims of this thesis are to contribute to the

- development and adaptation of methods that can deal with the related mathematically difficulties,

- identification of spatiotemporal spread patterns and their relation to the underlying mechanisms,

- improvement of the understanding of invasion processes,

- elucidation of possible forecast and control methods and

- development of a unified theory of biological invasions.

There are numerous examples of biological invasions, see for instance the classical book by Elton (1958) and, more recently, Hengeveld (1989); Drake et al. (1989); Williamson (1996); Shigesada and Kawasaki (1997); Mack et al. (2000); Pimentel (2002); Courchamp et al. (2003); Kowarik (2003); Capdevila-Argüelles and Zilletti (2005); Jeschke and Strayer (2005) and Snow et al. (2005). Why does this thesis concentrate on the role of infectious agents within bioinvasions? Infectious diseases are themselves prominent examples of invasions and continue to (re-)emerge even in modern times. The recent SARS outbreak clearly underlines the relevance and the need to understand the underlying mechanisms. Other recent epidemics such as AIDS, tuberculosis and malaria are responsible for the death of estimated 3 million, 2 million and 1 million people per year, respectively. Epidemics in history wiped out a quarter of the European population (Black Death, 14th century), half of the indigeneous population of the Aztecs (smallpox, 15th century) and even 20 million people in Europe in 1918/19 (Spanish flu). Still, also today's influenza strains are expected to cause severe epidemics. Reported cases of infectious diseases in animal populations, e.g. BSE, foot-and-mouth diseases, classic swine fever and avian flu, do not only threaten our live-stock – there is also an almost instantaneous response at the stock markets. Moreover, the consequences for human health are still not fully understood.

Two different fields of applications will be addressed, one being situated in a terrestrial ecosystem and one in an aquatic ecosystem. The first one deals with the spatiotemporal spread of an infectious disease in a mammal population, namely the Feline Immunodeficiency Virus (FIV) within populations of domestic cat (*Felis catus*, L.). The cat host population serves as a very suitable example of an harmful invader, because of which the impact of FIV on the invasion process is investigated as well. Recent studies indicate that pathogens and parasites play a major role in invasion biology (e.g. Keane and Crawley, 2002; Torchin et al., 2002; Mitchell and Power, 2003; Torchin et al., 2003; Lee and Klasing, 2004; Prenter et al., 2004).

The second case study deals with models of marine food chains, incorporating viral infection, diffusion and environmental noise. Just as in the previous example, the understanding of the role of infection is still at the beginning, but there is increasing evidence that marine viruses may exert control of oceanic life (e.g. Fuhrman, 1999). The focal phytoplankton-zooplankton community forms the basis for all other food webs in the sea and naturally occupies a central position in not only the marine but also in the global ecosystem functioning. Thus, the understanding of how viral infections may influence the plankton distribution and algal blooms is of considerable interest.

1.2   Outline

This thesis consists of two parts. In the first one, a two-component model for the spatiotemporal spread of an infectious disease (FIV) within cat populations is investigated. The second one deals with viral infections of phytoplankton in a prey-predator community, giving rise to three-component models.

First, it is focused on the spatial spread of an infectious disease in a terrestrial ecosystem. Therefore, a model for the circulation of a retrovirus within mammals, namely the Feline Immunodeficiency Virus (FIV) in domestic cat populations, is extended to a reaction-diffusion system, in which the diffusion describes the random dispersal of the cats (cf. Skellam, 1951; Okubo, 1980; Okubo et al., 1989). FIV induces what is called "cat-AIDS", i.e., infected animals suffer loss of their immune defence. Since the time to disease-induced death is rather long, the host population reproduction has to be taken into account. This has been done by assuming logistic growth (Courchamp et al., 1995a).

There are biological reasons that might induce a strong Allee effect, i.e. negative growth for small densities. For instance, such a growth rate can be caused by difficulties in finding mating partners, genetic inbreeding, demographic stochasticity or a reduction in cooperative interactions, see Dennis (1989); Courchamp et al. (1999a); Stephens et al. (1999) and Stephens and Sutherland (1999). Therefore, the FIV model is extended to incorporate a generalized strong Allee effect. The impact of an Allee effect in models of infectious diseases is largely unexplored, probably due to the fact that the historical development started with models assuming a constant population size (e.g. Kermack and McKendrick, 1927). For a more extensive introduction into epidemiological modelling, see the corresponding mini-review 1.4 below.

As it turns out, the joint interplay of the disease and the Allee effect has considerable impact upon both the host population and the transmission dynamics. Host population extinction can be observed as well as endemics, epidemics and the eradication of the disease. Moreover, the Allee effect induces bistability. Additionally accounting for spatial diffusion, travelling infection waves emerge when the disease is introduced in an established host population. However, if the inital conditions are finite for both the disease and the host population, i.e., the host itself invades empty space, then spatiotemporal dynamics such as travelling pulses and spatially restricted epidemics can be observed. Moreover, the effect of spatial heterogeneity is exemplarily investigated.

Computer simulations show that the disease has the potential to reverse the invasion front of the host population. This is of considerable interest from a management point of view (cf. Fagan et al., 2002). FIV has been proposed as a biocontrol agent (Courchamp and Sugihara, 1999; Courchamp and Cornell, 2000), because cats belong to the worst invasive species. In particular, they threaten native bird species on oceanic islands where the cats have been introduced as alien species, e.g. in order to prey upon rodents. From this point of view, the invading host population, i.e. the cat, is a pest which is aimed to be controlled by the introduction of FIV. The second chapter in the first part of the thesis is thus devoted to an analytical investigation by means of travelling wave approaches, in oder to derive conditions when the host invasion can be reversed, stopped or slowed down. The eventual extinction of the invading host might explain the phenomenon of invasion crashes, i.e., successfully invading populations suddenly collapse. Pathogens and parasites are frequently invoked to be responsible, but evidence was thus far lacking (Simberloff and Gibbons, 2004). Recessive invasion fronts are also referred to as *waves of extinction*, because the invasion would be stable in the absence of the disease. Similar dynamics have been observed in models of sterile insect release (Lewis and van den Driessche, 1993) and prey-predator communities (Owen and Lewis, 2001; Petrovskii et al., 2005c).

The second part deals with a series of models which are based upon a prey-predator system of phytoplankton-zooplankton interaction. Models of this type have been successful in understanding algal blooms and plankton patchiness, i.e., the spatially highly inhomogeneous distribution of plankton species, see for instance Medvinsky et al. (2002), Malchow et al. (2003) as well as the mini-review 1.6 below. Phytoplankton plays a key role in ecosystem functioning. It is the basis for all food chains and webs in the sea. Moreover, it produces half of the oxygen needed by mankind and absorbs half of the carbondioxide that may otherwise contribute to global warming. Suttle et al. (1990) have shown that marine viruses can infect bacteria and phytoplankton. There is increasing evidence, that viral infections might play a major role in aquatic ecosystems (cf. Fuhrman, 1999), but modelling approaches are just at the beginning (see the related mini-review 1.7).

The first chapter in the second part starts with focusing on lysogenic infection, i.e., viruses integrate their genome into the host cell's genome. When the host cell reproduces and duplicates its genome, the viral genome is duplicated, too. For modelling purpose, the phytoplankton is split into a susceptible and an infected part, both of which grow logistically with a common carrying capacity. Zooplankton is a predator grazing on both susceptible and infected phytoplankton. The local analysis yields a number of stationary and/or oscillatory regimes. The spatiotemporal behaviour is modelled as well. The diffusion now describes the eddy diffusion (e.g. Okubo, 1980; Okubo and Levin, 2001). A rich variety of dynamics can be observed, including spiral waves, dynamical stabilization of locally unstable equilibira as well as the "trapping" of infected phytoplankton. The latter occurs for initial conditions with finite patches of susceptible phytoplankton ahead of infected phytoplankton ahead of zooplankton.

Furthermore, the impact of environmental stochasticity is taken into account, which is modelled as multiplicative noise with stochastic partial differential equations (García-Ojalvo and Sancho, 1999; Allen et al., 2003; Anishenko et al., 2003). Their numerical treatment requires sophisticated algorithms. By way of common experience, the following choice of methods proved to be successful. The diffusion terms are integrated using the semi-implicit Peaceman-Rachford alternating direction scheme (Peaceman and Rachford Jr., 1955; Thomas, 1995). For the interactions and the Stratonovich integral of the noise terms, the explicit Euler-Maruyama scheme is applied (Maruyama, 1955; Higham, 2001; Kloeden et al., 2002). It turns out that noise can enhance the survival and spread of susceptibles and infected when they respectively would go extinct in a deterministic setting. Moreover, the noise also enhances the spread of infected when they would be "trapped" in a deterministic environment.

While the first chapter of this part models predator grazing with functional reponse of Holling-type II, the next chapter assumes a Holling-type III functional response. The local model thus becomes excitable, i.e., small perturbations can lead to a long-lasting course through the phase space before relaxation. This has been introduced by Truscott and Brindley (1994b) to model phytoplankton blooms. In the stochastic spatially extended model, the noise can induce local blooms of both susceptible and infected phytoplankton.

The next chapter returns to Holling-type II predation and provides a mathematical and numerical analysis of the local model for both lysogenic and lytic infection. Lysis leads to the destruction of the host cell as a consequence of infection. Particularly, the existence as well as the sudden disappearance of a mathematically interesting continuum of nontrivial solutions are shown. The equilibria on this continuum line are degenerated and undergo a fold-Hopf (zero-pair) bifurcation (Kuznetsov, 1995; Nicolis, 1995). The exact conditions for the existence of this solution continuum are probably nearly never met in nature. Instead, a strange periodic attractor emerges, which stabilizes itself after repeated torus-like oscillations.

Finally, the last chapter incorporates a transition from lysogeny to lysis into the stochastic spatiotemporal model. In nature, the lysogenic replication cycle is rather sensitive to environmental fluctuations and switches to the lytic cycle. This switch is implemented both deterministically and stochastically. While in the deterministic case a complex wavy structure appears that forms almost pinned pairs of spirals, the stochastic model disturbs the "pins" and blurs the artifically looking deterministic patterns. Thus, the arising spatiotemporal patterns look more realistic. This has also been demonstrated by Malchow et al. (2004a).

In the remaining sections of this introductory chapter, mini-reviews on the related scientific fields are given. The chapters of the following two parts are presented in a uniform style and each of the chapter has an own list of cited references. At the end of this thesis, a cohesive bibliography is given with all references from the chapters as well as the introduction and the mini-reviews.

1.3   Mini-review: Models of spatial spread and biological invasions

Biological invasions occur when species spread into environments which they have not previously occupied and become established there. Roughly, three different stages are usually distinguished: (i) the introduction of alien species into a localized domain ('local invasion'), (ii) the growth and establishment of the alien species if the local conditions are favourable and (iii) the large-scale spread ('global invasion').

The last stage may lead to dramatic changes in the community structure of ecosystems. The spatial spread of species is thus of extreme importance. Most mathematical models have focused on the long-term convergence of invasion dynamics to stage (iii). The basic "ingredients" of standard invasion models are population growth and spatial spread. The seminal paper by Skellam (1951) substantially initiated theoretical studies. Investigating the invasions of muskrats (*Ondatra zibethica*) which had escaped from a farm near Prague, he found that their range expansion increases linearly with time, i.e., the muskrats spread at a constant rate. He proposed a reaction-diffusion model with exponential growth, which successfully reproduces the observed spread pattern. The Skellam model thus became a paradigm in modelling biological invasion.

Actually, the Skellam model is a special case of the Fisher model (1937).

Fisher proposed a reaction-diffusion model with logistic growth as a deterministic version of a stochastic model for the spatial spread of an advantageous gene in a population. This model and its travelling frontal wave solutions have been intensively studied, in particular by Kolmogorov et al. (1937) who published their results independently in the same year. A traveling frontal wave is a solution of partial differential equation(s) that propagates with constant speed and does not change its shape. To be more precise, speed means the "asymptotic speed of propagation" (Aronson and Weinberger, 1975), cf. also Hadeler and Rothe (1975); Aronson and Weinberger (1978); Weinberger (1978); Fife (1979); Britton (1986) and Diekmann and Heesterbeek (2000). It turns out that both the Skellam and the Fisher wave move with the same asymptotic speed which is determined by the product of the diffusivity and the intrinsic growth rate at zero density. This is because the density dependence of the growth rate becomes irrelevant at the leading edge of the front which is typical for so-called "pulled" fronts. In contrast, when the growth rate exhibits a strong Allee effect, i.e., it is negative at low densities, then the wave becomes "pushed", because diffusion has to overbalance the decrease in density at the leading edge at the front where the density is small (Lewis and Kareiva, 1993). Moreover, invasion cannot take place unless it exceeds a critical area, cf. Nitzan et al. (1974); Malchow and Schimansky-Geier (1985); Lewis and Kareiva (1993) and van Saarlos (2003). It should be noted that already Luther (1906) found the speed of travelling waves in a reaction-diffucion model with exponential growth for chemical reactions of first order.

Diffusion models generally assume that individuals completely randomly move, but that their spread can be macroscopically described by the diffusion equation. The application of reaction-diffusion models ranges from ecology and epidemiology to bacterial growth, population genetics as well as to the spread of innovations or human cultures. Within ecology, reaction-diffusion models are widespread for mammals, insects, birds and plants, cf. particularly Shigesada and Kawasaki (1997) and Murray et al. (1986); Lubina and Levin (1988); Okubo et al. (1989); Yachi et al. (1989); Andow et al. (1990); Liebhold et al. (1992); Veit and Lewis (1996); Abramson et al. (2003). More textbook and article references include Okubo (1980); Edelstein-Keshet (1988); van den Bosch et al. (1990); Andow et al. (1990); Mollison (1991); Andow et al. (1993); Holmes et al. (1994); Shigesada and Kawasaki (1997); Kot (2001); Okubo and Levin (2001) and Cantrell and Cosner (2003).

However, there are biological invasion phenomena to which the Fisher model cannot be applied. Obviously, the spread is governed by other/additional factors which produce larger and smaller spread rates or even completely different spread patterns (cf. Hastings et al., 2005). In particular, spread rates have been observed which are biphasic or which seem to increase exponentially with time. This can be attributed to modes of spread different than diffusion. Two model approaches shall be mentioned here which can deal with this aspect. First, integro-difference equations are flexible enough to incorporate different dispersal kernels which describe the species' dispersal probabilities with respect to the dispersal distance. Thus, they are appropriate to model long-distance dispersal. If the kernel is Gaussian, one obtains diffusive spread. Weinberger (1982) showed that models approach a constant wave speed, provided that the kernel is exponentially bounded. More leptokurtic kernels typically yield a larger speed (Kot et al., 1996). 'Fat-tailed' dispersal kernels (kernels with tails that are not exponentially bounded) result in rapid, patchy and accelerating expansion that does not approach a constant velocity (Mollison, 1972, 1977; Kot et al., 1996). Second, invasion can take place from several locations/foci or an initial founder population can give rise to new colonies by long-distance migrants. Shigesada et al. (1995) and Shigesada and Kawasaki (1997, 2002) proposed 'scattered colony' and 'coalescing colony' models for this type of invasion which is better known as 'stratified diffusion' (Hengeveld, 1989) – or as 'hierarchical diffusion' in epidemiology.

Furthermore, recent model extensions have taken into account stochasticity (Lewis, 1997, 2000; Lewis and Pacala, 2000), resource availability, spatial heterogeneity (Shigesada et al., 1986; Murray, 2003), environmental borders (Keitt et al., 2001), inital conditions (Sherratt et al., 1995, 1997; Petrovskii and Malchow, 1999, 2001), predation (Petrovskii et al., 2002a, 2005a,c), competition (Okubo et al., 1989; Hart and Gardner, 1997), the Allee effect (Lewis and Kareiva, 1993; Wang and Kot, 2001; Wang et al., 2002; Petrovskii and Li, 2003; Petrovskii et al., 2005c), evolutionary changes (Lambrinos, 2004), large-scale phenomena such as weather conditions or (long-range) dispersal/transport effects (Hengeveld, 1989; Williamson, 1996; Shigesada and Kawasaki, 1997; Clark et al., 2001b; Petrovskii and Li, 2003).

In a series of papers, Petrovskii and co-workers provide a number of exact solutions for specific invasion problems. This is remarkable, because population models are nonlinear and solutions are usually only known for simple particular cases (e.g. Kierstead and Slobodkin, 1953). Petrovskii and Shigesada (2001) investigate a generalized Fisher model and especially focus on early invasion stages with increasing rates of expansion. They also relate their solution to the problem of critical aggregation, i.e., whether the introduced exotic species 'globally' invades or retreats, by applying the "practical stability concept" (i.e., a small threshold below which the population is assumed to go extinct). A reaction-diffusion-advection model with strong Allee effect is investigated by Petrovskii and Li (2003), where the density-(in)dependent migration accounts for environmental factors such as passive transport and biological motility mechanisms, respectively. The joint interplay of random dispersal and migration leads to various possibilities of either species invasion or retreat. Petrovskii et al. (2005c) provide an exact solution of a reaction-diffusion system describing prey-predator interactions with an Allee effect in the prey species and a nonlinear predator per-capita mortality rate. They derive exact conditions for front reversal.

Sherratt et al. (1995, 1997) have shown the emergence of spatiotemporal chaos in the wake of invasion fronts. Petrovskii and Malchow (1999, 2000, 2001) extended these results by demonstrating that irregular dynamics can stem from slightly inhomogeneous initial distributions. Invasion does not necessarily take place via regular population fronts. An Allee effect along with particular initial conditions in a two-species model can induce the break-up of travelling frontal waves and lead to patchy spread. This is a purely deterministic mechanism and can be attributed to two spatial dimensions of the system. The patchy spread usually takes place at the 'edge of extinction' (Petrovskii et al., 2002a,b, 2005a,b).

For recent reviews of models of bioinvasions, see Fagan et al. (2002); Case et al. (2005); Hastings et al. (2005); Holt et al. (2005) and Petrovskii et al. (2005b). Some of these factors have been shown to generate patchy spread, dynamic stabilization of an unstable equilibrium (Petrovskii and Malchow, 2000;

Malchow and Petrovskii, 2002), chaos in the wake of invasions and the slow-down or retreat of invasion fronts.

This thesis particularly focuses on the joint interplay of various of these factors. This has been shown to make possible the following patterns of spatiotemporal spread:

- trapping in an epidemiologically extended prey-predator model (Malchow et al., 2004b, 2005; Petrovskii et al., 2005b),

- noise-induced spread and survival of infected (or susceptibles) as well as patchy spread in an epidemiological model with stochasticity (Malchow et al., 2004b, 2005; Petrovskii et al., 2005b),

- patchy spread and spatially restricted invasions in a deterministic model with specific initial conditions and an Allee effect either in the prey species of a prey-predator system (Petrovskii et al., 2002b) or in the host population of an infectious disease (Petrovskii et al., 2005b),

- blocking or reversal of invasion fronts in models with specific initial conditions and an Allee effect either in the prey species of a prey-predator system (Owen and Lewis, 2001; Petrovskii et al., 2005c) or in the host population of an infectious disease (Hilker et al., 2005). In the latter case, temporally and spatially restricted invasions are possible as well (Hilker et al., accepted).

1.4   Mini-review: Mathematical epidemiology

The transmission dynamics of infectious diseases is one of the oldest topics in mathematical biology. Already in 1760, Daniel Bernoulli provided by using a single nonlinear ordinary differential equation the first known mathematical result of epidemiology, that is the defence of the practice of inoculation against smallpox. Hamer (1906), modelling the recurrence of measles epidemics with discrete-time models, was probably the first to assume that the incidence (number of new cases per unit time) depends on the product of the densities of the susceptibles and infected – what is well-known as the law of mass action e.g. from reaction kinetics or Lotka-Volterra systems (Lotka, 1925; Volterra, 1926). Ross (1911) developed differential equation models for malaria as a host-vector disease. The first attempt to account for births and deaths within an epidemic model is probably due to Soper (1929), with the aim to explain the periodicity in measles infections.

Most influential in the compartmental modelling of infectious diseases has been the work of Kermack and McKendrick (1927, 1932, 1933). Compartmental modelling means that the host population is divided into a susceptible part, traditionally denoted by S, an infectious/infective/infected part (I) and a removed part (R), the latter comprising individuals that died, are in quarantine or that gained immunity with recovery. This gives rise to the so-called SIR-model. Many other compartments have been introduced, e.g. an exposed but not infectious yet class (E), thus giving rise to SEIR models. The series of papers by Kermack and McKendrick attended to explain the temporal dynamics of a disease in a homogeneous isolated population and gave fundamental ideas explaining why a disease may be epidemic, endemic or not, or what are the main factors controlling the disease in the population. They also show that diseases can be a population control factor. Very well-known is their threshold theorem which states that the number of susceptibles must exceed a critical value to stimulate an epidemic outbreak. It is based on the basic reproductive number R_0, which is the average number of secondary infections produced when one infected individual is introduced into a host population where everyone is susceptible. If $R_0 < 1$ the infection dies out, if $R_0 > 1$ there is an epidemic.

In the last decades, the amount of works in this area has exploded. The first edition of Bailey's book (1957) was an important landmark. Influential work is also due to Anderson and May, e.g., Anderson and May (1979); May and Anderson (1979); Anderson et al. (1981) and Anderson and May (1991). Generally, different aspects are dealt with in the literature, from human health to environmental assessment. Moreover, modelling disease transmission is a subject where mathematics is deeply involved, from simple model analyses to the development of unifying methods to deal with large classes of models. Recent models have involved aspects such as vaccination, partial or temporary immunity, quarantine, stages of infection, vertical transmission, disease-reduced fertility, seasonality, evolutionary aspects, age structure, spatial spread, social and sexual mixing groups, multiple disease strains, multiple/reservoir hosts, multiple diseases in one host, variable infectiousness, stochasticity and disease vectors. Reviews, textbooks and collections include Busenberg and Cooke (1993); Capasso (1993); Hethcote (1994); Diekmann and Heesterbeek (2000); Hethcote (2000); Brauer and Castillo-Chavez (2001); Castillo-Chavez et al. (2002a,b); Murray (2002) and Thieme (2003).

Within this thesis, special emphasis will be on the following aspects.

- When the duration of the disease is long in comparison with the life span of the host and there is a disease-related mortality, population dynamics have to be taken into account with a variable population size (Anderson and May, 1979; Anderson et al., 1981; Brauer, 1990; Busenberg and van den Driessche, 1990; Greenhalgh, 1990; Pugliese, 1990; Diekmann and Kretzschmar, 1991; Gao and Hethcote, 1992; Greenhalgh, 1992; Mena-Lorca and Hethcote, 1992; Zhou and Hethcote, 1994; Greenhalgh and Das, 1995; Hilker et al., accepted).

- The circulation of diseases is also affected by the spatial spread of the host population. Hence, there is a need to incorporate the spatiotemporal interactions (Brownlee, 1911; Kendall, 1965; Noble, 1974; Mollison, 1977; Murdoch and Bence, 1987; van den Bosch et al., 1990; Mollison, 1991; Meade, 1992; Sato et al., 1994; Johansen, 1996; Scherm, 1996; Holmes, 1997; Shigesada and Kawasaki, 1997; Hilker et al., accepted). Recently, models based on network and graph theory as well as multi-patch and metapopulation models are increasingly developed (e.g., Barbour and Mollison, 1990; Diekmann et al., 1990; Andersson, 1998; Kleczkowski and Grenfell, 1999; Moore and Newman, 2000; May and Lloyd, 2001; Pastor-Satorras and Vespignani, 2001; Arrigoni and Pugliese, 2002; Park et al., 2002; Arino and van den Driessche, 2003; Neal, 2003; Barthélemy et al., 2005).

- Diseases are transmitted in various ways. For instance, childhood diseases such as measels will potentially be transmitted to as many individuals as in the susceptible population, while sexually transmitted diseases will be spread among a more or less constant number of contacts independent of the population size. The first example with a linearly increasing contact number gives rise to the mass action principle of the classical disease models, while the latter example of a constant contact number is referred to as standard incidence, proportionate mixing or frequence-dependent transmission (Hethcote, 1976; Nold, 1980; Getz and Pickering, 1983; Busenberg and van den Driessche, 1990; McCallum et al., 2001). Moreover, many other transmission functions / forces of infections have been proposed, e.g., saturation-type functions (Diekmann and Kretzschmar, 1991; Ruan and Wang, 2003; Zhang and Ma, 2003; Kribs-Zaleta, 2004; Moghadas, 2004), power-law-type functions (Liu et al., 1986, 1987; Hethcote et al., 1989; Hethcote and van den Driessche, 1991; Thieme, 1992; Derrick and van den Driessche, 1993; Greenhalgh and Das, 1995; Li and Muldowney, 1995; Greenhalgh, 1997; van den Driessche and Watmough, 2000; Derrick and van den Driessche, 2003; van den Driessche and Watmough, 2003; Ruan and Wang, 2003; Alexander and Moghadas, 2004; Korobeinikov and Maini, 2004), affine functions (Greenhalgh and Das, 1995; van den Driessche and Watmough, 2000, 2003), negative binomial functions, asymptotic contact and transmission rates (McCallum et al., 2001), indirect transmission functions (Ghosh et al., 2005) and generalized (Li and Wang, 2002; Li and Zhen, 2005; Wang and Li, 2005) as well as continuum functions (Antonovics et al., 1995; McCallum, 2000; Boots and Sasaki, 2003).

- One of the fundamental aspects in mathematical epidemiology concerns the mechanisms of periodicity, that is, in biological words, the recurrence of epidemics. For quite a long time, mathematical models have dealt with damped oscillations, i.e., an endemic equilibrium which is a stable focus (cf. Soper, 1929; Wilson and Worcester, 1945a,b; Bailey, 1975). In the 1970s and 1980s, several factors have been identified to generate sustained oscillations. They are reviewed in Hethcote et al. (1981b) and Hethcote (1989). Nowadays, many other mechanisms have been found. They can be summarized as follows: seasonal forcing (Hethcote, 1973; London and Yorke, 1973), delays in the removed class and the number of epidemiological classes (Gripenberg, 1980; Hethcote et al., 1981a; Stech and Williams, 1981; Diekmann and Montijn, 1982; Smith, 1986), nonlinear transmission rates (May and Anderson, 1978; Liu et al., 1986, 1987; Diekmann and Kretzschmar, 1991; Derrick and van den Driessche, 1993) and variable population size (Anderson et al., 1981; Pugliese, 1991; Mena-Lorca and Hethcote, 1992; Gao et al., 1995; Greenhalgh, 1997), stochastic factors (cf. May and Anderson, 1979), multiple diseases or multiple strains (Andreasen et al., 1997; Lin et al., 1999; Dawes and Gog, 2002; Levin et al., 2004), age structure (Enderle, 1980; Schenzle, 1984; Kubo and Langlais, 1991; Thieme and Castillo-Chavez, 1993a,b; Milner and Pugliese, 1999; Martcheva and Castillo-Chavez, 2003), spatio-temporal interactions (Johansen, 1996) and core groups (Kribs-Zaleta, 1999).

- The spread of an infectious disease can also be affected by interferences with ecological aspects, e.g, by community interactions. Hadeler and Freedman (1989) as well as Freedman and Moson (1990) were probably the first to model parasitic infection in a prey-predator system. The joint interplay of infectious diseases and predation has then been investigated by Venturino (1995); Mukherjee (1998); Chattopadhyay and Arino (1999); Xiao and Chen (2001a,b); Chattopadhyay et al. (2003a); Choo et al. (2003); Mukherjee (2003); Zhdanov (2003) and Hethcote et al. (2004). The mini-review in section 1.7 surveys the special case of viral infections in planktonic food chains (Beltrami and Carroll, 1994; Chattopadhyay and Pal, 2002; Chattopadhyay et al., 2003b; Malchow et al., 2004b; Singh et al., 2004; Malchow et al., 2004b, 2005). The role of diseases in competition models was studied by Bowers and Turner (1997); Frantzen (2000); Venturino (2001) as well as Wodarz and Sasaki (2004). Some authors used the term "eco-epidemiological" models to describe the merging of ecological and epidemiological approaches, but it should be noted that "eco-epidemiology" has also other roots (e.g., Susser and Susser, 1996; Susser, 2004).

1.5 Mini-review: FIV in cat populations

The Feline Immunodeficiency Virus (FIV) is a retrovirus ("slow virus") that causes an immunodeficiency condition in cats, i.e., damage to the immune system. It is the same subfamily (lentivirus) that causes human AIDS (Acquired Immunodeficiency Syndrome), but the FIV infection is not equally fatal. Usually, cats still have quite a long life expectation. FIV has been found worldwide in a large number of felid species. The focus of this thesis will be restricted to FIV within populations of the domestic cat (*Felis catus*, L.), for which the prevalence of infection varies from approximately 1 to 20 %.

Transmission of FIV occurs almost exclusively through bite wounds (Yamamoto et al., 1989; Ueland and Nesse, 1992; Bendinelli et al., 1995; Courchamp et al., 1998), e.g., when cats fight for females monopolization or territory defence. Because of this, the disease is more common in roaming and dominant male cats (e.g. Courchamp et al., 1995b, 1998, 2000b). Another difference to HIV (Human Immunodeficiency Virus) is that the virus is not transmitted to offspring, i.e., there is no vertical transmission (Ueland and Nesse, 1992; Sellon et al., 1994; O'Neil et al., 1996; Rogers and Hoover, 1998). There is no evidence that transmission to humans can occur. For an epidemiological review, see Courchamp and Pontier (1994).

The clinical course of FIV usually goes through five stages. At first, exposed cats suffer an acute stage lasting 2 to 4 weeks (Yamamoto et al., 1988; Brandon, 1995); most of the exposed animals become lifelong carriers of the virus. Then there is an asymptomatic carrier phase lasting months to years. During this long stage, the cat is in good health conditions and its behaviour seems to be normal. The last three stages are persistent lymphadenopathy, ARC (for AIDS Related Complex) and AIDS. They are characterized by miscellaneous disorder, chronic infections, loss of immune defence and eventually death (Sparger, 1993). During this short period, cats are too weak to participate in social life, competition or mating. There is neither natural immunization nor natural recovery (Zenger, 1992).

For modelling purposes, it can be assumed that there is only one stage after infection combining the acute, asymptomatic, persistent lymphadenopathy,

ARC and AIDS phases (Courchamp et al., 1995a). This is reasonable because the asymptomatic phase is usually much longer than the other ones. Thus, there is a susceptible and an infected compartment of the host population with no vertical transmission, no recovery from the disease and a small disease-related mortality. Since the time being infected can be long in comparison with the lifespan, the host population growth is taken into account. Furthermore, the standard incidence (also called proportionate mixing) is assumed because of a constant number of contacts with bites (Courchamp et al., 1995a). This is reasonable for populations in rural/suburban and non-anthropized habitats with cat densities from 100 to 1000 and smaller than 10 individuals per km^2, respectively, while in urban habitats with more than 1000 individuals per km^2 or rural/suburban habitats with 10-100 individuals per km^2 mass action transmission seems to be appropriate (Fromont et al., 1998).

The model by Courchamp et al. (1995a) has been the first one describing the dynamics of a retrovirus transmission within populations of mammals (except the special case of humans). It is a simple SI model and has been extended to an SEI (Langlais and Suppo, 2000) and an SIS model (Fitzgibbon and Langlais, 2003; Ainseba et al., 2002) from more mathematical points of view. The latter reference is based on a reaction-advection-diffusion approach, thus being able to cope with spatially heterogeneous population dynamics. Models of spatial spread can also be found for the Feline Leukaemia Virus (FeLV), which is another retrovirus in cats. In Fitzgibbon et al. (2001), the spread of FeLV through highly heterogeneous habitats has been mathematically analyzed and in Fromont et al. (2003) two host populations with density-dependent dynamics have been connected together as well as to the surrounding environment by dispersal.

As FIV may occur simultaneously with FeLV, approaches have been developed to model the circulation of more than one microparasite in a single host population (Courchamp et al., 1995c; Allen et al., 2003). To investigate the influence of different behavioral traits of cat individuals, the fast game theory has been coupled to slow population dynamics of cats (Auger and Pontier, 1998; Pontier et al., 2000) and applied to a FIV model as well (Bahi-Jaber et al., 2003). Different behavioural traits and stochastic fluctuations in individual parameters have also been investigated in Bahi-Jaber et al. (2003).

Domestic cats threaten wildlife in various ways. They are not only acting as disease vector of feline diseases such as FIV or FeLV. Cats are a vector for rabies and SARS, too, which they can spread to humans and animals (Martina et al., 2003). Furthermore, introduced cats pose a severe threat to native birds on oceanic islands (Courchamp et al., 2003). Hence, they are regarded as one of the worst invasive alien species. FIV has been proposed as a biocontrol agent (Courchamp and Sugihara, 1999; Courchamp and Cornell, 2000), but its potential impacts have to be carefully assessed, cf. Courchamp et al. (1999b,c, 2000a).

This thesis will be concerned with

- extending the existing FIV model with logistic host population growth by Courchamp et al. (1995a) to incorporate generalized fertility and mortality functions; in particular, a strong Allee effect will be investigated (Hilker et al., accepted),

- the spatiotemporal spread of FIV in a reaction-diffusion model both from a numerical and an analytical point of view (Hilker et al., 2005, accepted),

- the impact of spatial heterogeneity which is exemplarily modelled by assuming spatially varying parameters, e.g., various carrying capacities in more or less favourable habitats and

- the potential of FIV to control the invasion of its host (Hilker et al., 2005).

1.6 Mini-review: Pattern formation in plankton dynamics

Plankton are small organisms that inhabit the water column of of the ocean, seas, and bodies of freshwater. The name comes from the Greek term *planktos* = made to wander. Within the focus of this thesis are two functional groups of plankton, namely phytoplankton and zooplankton. Phytoplankton (from *phyton* = plant) are microscopic plants near the water surface. They drive all marine communities and the life within them. Their photosynthetic growth generates half of the oxygen needed by mankind. Furthermore, phytoplankton absorbs half of the carbondioxide that may contribute to global warming as well as many other substances. Hence, phytoplankton is one of the main factors controlling the further development of the world's climate. Zooplankton (from *zoon* = animal) are the animals in plankton. Both herbivores and predators occur in marine plankton; the herbivores graze on phytoplankton and are eaten by the predators. Together, phyto- and zooplankton form the basis for all food chains and webs in the sea.

The spatial horizontal distribution of plankton in the natural marine environment is highly inhomogeneous. Plankton populations are patchy on temporal and all spatial scales. This phenomenon of plankton "patchiness" is affected by a number of factors which depend on the spatial scale. On a scale of dozens of kilometers and more, the plankton abundance is mainly controlled by the inhomogeneity of underlying hydrophysical fields like temperature and nutrients. On a scale of less than one hundred meters, plankton patchiness is controlled by turbulence. Furthermore, there is an intermediate scale, roughly from a hundred meters to a dozen kilometers, where the plankton distribution is uncorrelated with the environment and, thus, the nature of the spatial heterogeneity is essentially different. This distinction is usually considered as the prevailing of biological factors against hydrodynamics. An elementary introduction into the subject of plankton patchiness and the bibliography is given by Malchow et al. (2003) from a modelling point of view.

Generally, the growth, competition, grazing and propagation of plankton communities can be described by partial differential equations of reaction-diffusion type. Further aspects can be modelled as well, e.g., the effects of water movement by flow; seasonal effects by external periodic forcing; patchiness of nutrients for algae by heterogeneity; and the lag between the build-up of phyto- and zooplankton by delays. Exemplary references include Malchow and Shigesada (1994); Steffen et al. (1997); Medvinsky et al. (2001a,b) and Medvinsky et al. (2004).

Recent models of the prey-predator interaction of phytoplankton and zooplankton dynamics take into account zooplankton grazing with saturating functional response to phytoplankton abundance, which are also known from Monod or Michaelis-Menten models of enzyme kinetics (Michaelis and Menten, 1913; Monod and Jacob, 1961). These models can explain the well-known prey-predator oscillations as well as the oscillatory or monotonous relaxation to one of the possible multiple equilibria, e.g. Steele and Henderson (1981); Scheffer (1991a); Steele and Henderson (1992b,a); Malchow (1993); Pascual (1993); Truscott and Brindley (1994a,b) and Scheffer (1998). Excitable systems exhibit a long-lasting relaxation to the equilibrium after a supercritical external perturbation like a sudden temperature increase or nutrient inflow. They are especially interesting, since they are capable of modelling red or brown tides (Beltrami, 1989, 1996; Truscott and Brindley, 1994a,b).

The spectrum of spatial and spatio-temporal patterns includes regular and irregular oscillations, propagating fronts, target patterns and spiral waves, pulses as well as stationary spatial patterns. Many mechanisms generating these patterns can also be found in the section about modelling biological invasions. Among them are, for instance, the Fisher and Skellam front waves (Luther, 1906; Fisher, 1937; Kolmogorov et al., 1937; Skellam, 1951), dynamical stabilization of unstable equilibria (Petrovskii and Malchow, 2000; Malchow and Petrovskii, 2002) and chaotic oscillations behind propagating diffusive fronts in prey-predator models with finite (Sherratt et al., 1995, 1997) or, more generally, slightly inhomogeneous intial distributions (Petrovskii and Malchow, 1999, 2000, 2001). When the population growth is governed by an Allee effect, the regular fronts break up and patchy spread can be observed (Petrovskii et al., 2002a,b, 2005b).

Segel and Jackson (1972) were the first to apply the concept of diffusive instabilities to a problem in population dynamics, namely the prey-predator interaction of phytoplankton and herbivorous copepods with higher herbivore motility. Diffusive instabilities and, thus, spatial structure can arise in a nonlinear system of at least two species when their diffusivities are essentially different. In this case, a supercritical perturbation destabilizes a spatially homogeneous stable stationary state. This scenario of spatial pattern formation is closely associated with the classical paper by Turing (1952) and suggested by Levin and Segel (1976) for a possible origin of planktonic patchiness. The generation and drift of planktonic Turing patches are described by Malchow (1993, 1994, 2000a). Turing patterns look rather artificial because of their symmetry. Malchow et al. (2004a,b) have investigated the influence of environmental noise and shown that it blurs the "artificial" patterns to let them look more realistic.

The destabilization of spatially uniform populations can also be attributed to differential-flow-induced instabilities (Rovinsky and Menzinger, 1992), which has been applied to plankton dynamics by Malchow (1995, 1996, 1998). Spatial patterns can emerge if the phytoplankton and the zooplankton move with different velocities but regardless of which one is faster. This mechanism is more general than Turing's diffusive instabilities, since the latter is restricted to strong conditions on the diffusion coefficients.

Kierstead and Slobodkin (1953) were perhaps the first to think of the critical size problem for plankton patches. They proposed a reaction-diffusion model of a single population with exponential growth, which is nowadays known as KISS model. In contrast to the Skellam model (1951), the spatial domain is not infinite, but restricted by homogeneous Dirichlet boundary conditions. They derived a critical length, which is referred to as the KISS size. If the plankton patch is shorter than this length, the phytoplankton population collapses. If it is longer than this length, a bloom occurs. Other mechanisms inducing a critical patch size are reviewed in the section about biological invasions.

That the spatial dimension of the plankton community provides routes to chaotic dynamics has first been shown by Pascual (1993). The spatial distribution of nutrients along a linear gradient can induce spatiotemporal chaos. In fact, there appear stronger and stronger indications in favour of the existence of chaos in population dynamics (Hanski et al., 1993; Costantino et al., 1995; Dennis et al., 1995; Costantino et al., 1997; Turchin and Ellner, 2000; Becks et al., 2005), see also Cushing et al. (2003) and Turchin (2003). Chaotic population dynamics essentially changes the approach to the system predictability, cf. Scheffer (1991b), and makes conceptual few-species models of as much use as many-species ones. Moreover, few-species models can sometimes be even more instructive since they take into account only the principal features of the community functioning. Therefore, mainly minimal models are described here.

Furthermore, it should be noted that this section has focused on reaction-diffusion models. In order to provide a more complete overview, which can be found in Malchow et al. (2003), one has to additionaly account for directed and joint or relative motion of the species. Also hybrid models of equation-based plankton dynamics and rule-based fish school dynamics have been investigated (Malchow et al., 2000, 2002). An extensive review on spatiotemporal pattern formation in plankton dynamics is given by Medvinsky et al. (2002). The books by Okubo (1980) and Okubo and Levin (2001) may be consulted, too, as well as, regarding general pattern formation, the textbook by Murray (2003).

This thesis introduces the role of viral infections into spatiotemporal pattern formation mechanisms within plankton dynamics (Malchow et al., 2004b, 2005). The next section briefly reviews infectious diseases in plankton and the related modelling approaches.

1.7   Mini-review: Viral infections in plankton communities

Viruses are evidently the most abundant entities in the sea, typically numbering ten billion per litre. Thus, the question concerning their role in oceanic life arises. E.g., may they possibly control aquatic food webs, phytoplankton blooms and algal biodiversity?

Though viruses have been ignored in microbial studies of the marine food web until one/two decades ago, an increasing number of reports gives evidence of the significant effect of viral agents on phytoplankton and bacteria as well as their biogeochemical and ecological effects, cf. the review by Fuhrman (1999). The presence of pathogenic viruses in phytoplankton communities is shown in Bergh et al. (1989); Suttle et al. (1990); Wilhelm and Suttle (1999); Tarutani et al. (2000) and Wommack and Colwell (2000), and virus-like particles are described in van Etten et al. (1991); Peduzzi and Weinbauer (1993); Reiser (1993) and Suttle et al. (1993). There is some evidence that viral infection might accelerate the termination of phytoplankton blooms (Tarutani et al., 2000; Jacquet et al., 2002; Gastrich et al., 2004). Viruses are held responsible for the collapse of *Emiliania huxleyi* blooms in mesocosms (Bratbak et al., 1995)

and in the North Sea (Brussard et al., 1996) and are shown to induce lysis of
Chrysochromulina (Suttle and Chan, 1993). Furthermore, because most viruses
are strain-specific, they can increase genetic diversity (Nagasaki and Yamaguchi,
1997). Nevertheless, despite the increasing number of reports, the role of viral
infection is still far from understood.

Viral infections of phytoplankton cells can be lysogenic or lytic. The understanding of the importance of lysogeny is just at the beginning (Wilcox and
Fuhrman, 1994; Jiang and Paul, 1998; McDaniel et al., 2002; Ortmann et al.,
2002). Contrary to lytic infections with destruction and without reproduction
of the host cell, lysogenic infections are a strategy whereby viruses integrate
their genome into the host's genome. As the host reproduces and duplicates its
genome, the viral genome reproduces, too.

Mathematical models of the dynamics of virally infected phytoplankton populations are rare as well. The already classical publication is by Beltrami and
Carroll (1994). They consider a phytoplankton-zooplankton prey-predator system and introduce an SI disease in the prey. They argue that the transmission
rate is not limited by the availability of cells because of the typically high algal
density. Therefore, they choose a frequency-dependent transmission function
(proportionate mixing) for which the contact rate is roughly proportional to the
infected phytoplankton population if the susceptible population is large compared to the infected one. The infected do not reproduce, but suffer cell lysis
with a constant per-capita rate. Moreover, the infected contribute to the carrying capacity of the logistic growth of the susceptibles. Zooplankton grazes on
both susceptibles and infected, modelled by simple Lotka-Volterra-type interaction.

Beretta and Kuang (1998) model the infection of marine bacteria by infective
viruses (bacteriophages). They explicitly take into account an equation for the
phages which are released by cell lysis. Mass action transmission is assumed with
the reasoning that usually many viruses (on average, more than ten) attack a
susceptible bacterium, but only one virus enters its head through the bacterial
membrane and can start its replication. The infected are assumed to be removed
by lysis before having the possibility of reproducing, but they still influence the
growth of the susceptibles towards the carrying capacity.

More recent work concerning viral infections of phytoplankton is by Chattopadhyay and co-workers (Chattopadhyay and Pal, 2002; Chattopadhyay et al.,
2002, 2003b; Singh et al., 2004) and Malchow et al. (2004b, 2005). The model
by Chattopadhyay and Pal (2002) draws upon Beltrami and Carroll (1994), but
differs in the following three aspects. First, it assumes intraspecific competition
of the zooplankton predator population. Second, mass action transmission is
assumed by arguing that the observed transmssion in the Greenwood experiments could just as well be explained with a transmission function that follows
the law of mass action and is not frequency-dependent (de Jong et al., 1995). In
the Greenwood experiments, Greenwood et al. (1936) quantified disease transmission among mice. Third, Chattopadhyay and Pal (2002) consider an SIS
disease in which infected cells can recover and become susceptible again. This
seems to be a very debatable assumption.

Chattopadhyay et al. (2002) model the effect of harvesting on all three
species. Basically, their system is similar to Beltrami and Carroll (1994), but
it assumes that the infected phytoplankton does not limit the growth of the
susceptible phytoplankton. Moreover, zooplankton grazing is assumed to be of

Holling-type II for the susceptible phytoplankton and of Lotka-Volterra-type for the infected phytoplankton. There cannot be found any explicit reasoning for this modelling, but it is mentioned that infection may modify the behaviour of prey species so that they become more vulnerable to predation. The cited examples are fish. However, what seems to be common opinion among biologists is that zooplankton grazers cannot distinguish between susceptible and infected zooplankton and do not graze on them with different rates.

Chattopadhyay et al. (2003b) consider a nutrient-phytoplankton system with virally infected phytoplankton. The disease transmission is frequency-dependent.

The model by Singh et al. (2004) is again similar to the one by Beltrami and Carroll (1994) – except the assumptions that transmission follows the law of mass action and that infected prey is more vulnerable to predation, cf. above.

Within this thesis, the following issues will be addressed:

- lysogenic infection and Holling-type II grazing (Malchow et al., 2004b),

- lysogenic infection and Holling-type III grazing (Malchow et al., 2005),

- lytic as well as lysogenic infection and Holling-type II grazing,

- switch from lysogenic to lytic infection (deterministically and stochastically) and Holling-type II grazing.

Additionally, the spatiotemporal dynamics and environmental stochasticity are taken into account by considering diffusion and multiplicative density-dependent noise, respectively (in all but the third issue). Zooplankton grazes equally on both susceptible and infected phytoplankton and does not suffer from grazing on infected phytoplankton (since virus infections are rather host-specific). In the case of lysogenic infection, susceptibles as well as infected reproduce, whereby the intrinsic growth rate of the susceptibles can be assumed to be greater than that of the infected. There is a joint carrying capacity to which both susceptibles and infected contribute, for instance by shading and need for space. In the case of lytic infection, the infected do not reproduce anymore and suffer an increased mortality. Competition becomes non-symmetric, because the infected still influence the growth of the susceptibles but not vice versa. The disease transmission is assumed to be frequency-dependent as in Beltrami and Carroll (1994).

Part I

Epidemic spread and host invasion

Chapter 2

A diffusive SI model with Allee effect and application to FIV

Frank M. Hilker[1], Michel Langlais[2], Sergei V. Petrovskii[3] and Horst Malchow[1]

[1] Institute of Environmental Systems Research, Department of Mathematics and Computer Science, University of Osnabrück, 49069 Osnabrück, Germany

[2] UMR CNRS 5466, Mathématiques Appliquées de Bordeaux, case 26, Université Victor Segalen Bordeaux 2, 146 rue Léo-Saignat, 33076 Bordeaux Cedex, France

[3] Shirshov Institute of Oceanology, Russian Academy of Science, Nakhimovsky Prospekt 36, Moscow 117218, Russia

June 9, 2005.
A slightly modified version has been accepted for publication in *Mathematical Biosciences*.

Abstract

A minimal reaction-diffusion model for the spatiotemporal spread of an infectious disease is considered. The model is motivated by the Feline Immunodeficiency Virus (FIV) which causes AIDS in cat populations. Because the infected period is long compared with the lifespan, the model incorporates the host population growth. Two different types are considered: logistic growth and growth with a strong Allee effect. In the model with logistic growth, the introduced disease propagates in form of a travelling infection wave with a constant asymptotic rate of spread. In the model with Allee effect the spatiotemporal dynamics are more complicated and the disease has considerable impact on the host population spread. Most importantly, there are waves of extinction, which arise when the disease is introduced in the wake of the invading host population or when spatial heterogeneities cause unfavourable habitats. These waves of extinction destabilize locally stable endemic coexistence states. Moreover, spatially restricted epidemics are possible as well as travelling infection pulses that correspond either to fatal epidemics with succeeding host population extinction or to epidemics with recovery of the host population. Generally, the Allee effect induces minimum viable population sizes and critical spatial lengths of the initial distribution. The local stability analysis yields bistability and the phenomenon of transient epidemics within the regime of disease-induced extinction. Limit cycle oscillations do not exist.

2.1   Introduction

There is an ongoing interest in the dynamics of infectious diseases and their spatiotemporal spread. Simple compartmental models have been used to understand the principal mechanisms governing disease transmission and have been extended to reaction-diffusion equations to estimate the asymptotic rate of spatial spread (Shigesada and Kawasaki, 1997; Diekmann and Heesterbeek, 2000; Murray, 2002, 2003; Cantrell and Cosner, 2003; Thieme, 2003). For diseases with long incubation and infectivity times, the host population's vital dynamics, i.e., birth and death rates, have to be taken into account. This has been shown to qualitatively change the system behaviour. While in the classic Kermack and McKendrick model (1927) the disease dies out or becomes endemic if the basic reproductive ratio is less or greater than one, respectively, the infection can also become endemic in models incorporating vital dynamics (Kermack and McKendrick, 1932; Brauer and Castillo-Chavez, 2001). If the disease additionally reduces the population size, one has to consider epidemiological models with varying population sizes (Anderson and May, 1979; Brauer, 1990; Busenberg and van den Driessche, 1990; Pugliese, 1990; Diekmann and Kretzschmar, 1991; Gao and Hethcote, 1992; Mena-Lorca and Hethcote, 1992; Zhou and Hethcote, 1994; Greenhalgh and Das, 1995). This may lead in particular cases to a destabilization of the endemic state and to sustained limit cycle oscillations, as has been numerically observed for the first time by Anderson et al. (1981).

The aim of this paper is to explore the consequences of different vital dynamics on the disease transmission as well as on the spatial spread. The starting point is an SI model with logistic growth, the standard incidence (also called proportionate mixing or frequency-dependent transmission) and no vertical transmission, which has been proposed and applied by Courchamp et al. (1995) to the circulation of the Feline Immunodeficiency Virus (FIV) within domestic cats (*Felis catus*, L.). FIV is a lentivirus which is structurally similar to the Human Immunodeficiency Virus (HIV) and induces feline AIDS (Acquired Immunodeficiency Syndrome) in cats.

The vital dynamics are generalized to be governed by a strong Allee effect. This can be caused by difficulties in finding mating partners at small densities, genetic inbreeding, demographic stochasticity or a reduction in cooperative interactions, see Dennis (1989); Courchamp et al. (1999); Stephens et al. (1999) and Stephens and Sutherland (1999). It should be noted, moreover, that the study of Allee dynamics is justified in its own rights, because this is largely lacking in the epidemiological literature, but see Hilker et al. (2005) and Petrovskii et al. (2005a).

The spatial propagation of diseases has been investigated in the literature mainly by way of travelling wave approaches and by approximating the asymp-totic rate of spread (Noble, 1974; Mollison, 1977; Murray et al., 1986; Okubo et al., 1989; Yachi et al., 1989; van den Bosch et al., 1990; Mollison, 1991; Dwyer, 1994; Caraco et al., 2002; Abramson et al., 2003; Rass and Radcliff, 2003). Two different scenarios will be considered in this study for both the models with logistic growth and Allee effect: On the one hand the spread of infection in a disease-free population which has settled down at carrying capacity in all the space and on the other hand the case that the host population itself still in-vades into empty space with the disease introduced in the wake of this invasion front. The latter case is of special interest, since the domestic cat is considered to be one of the worst invasive species and the introduction of FIV has been proposed as biological control method (Courchamp and Sugihara, 1999; Cour-champ and Cornell, 2000). Furthermore, the effect of spatial heterogeneity is briefly discussed by considering unfavourable habitats for the host population.

This paper is outlined as follows. In the next section, the basic SI model is de-scribed. Then, the results of a detailed local stability analysis of the model with a generalized strong Allee effect are presented. The spatiotemporal spread is studied by numerical simulations and travelling wave approaches in section 2.4. Finally, the results are discussed and related to similar work.

2.2   Model description

Let $P = P(t, \boldsymbol{x}) \geq 0$ be the density of the host population (in number of indiviuals per km^2) at time t (years) and spatial location $\boldsymbol{x}$ (in km). The fertility function $\beta(P) \geq 0$ and the mortality function $\mu(P) \geq 0$ are assumed to be density-dependent. Then the intrinsic per-capita growth rate is

$$g(P) = \beta(P) - \mu(P) \ .$$

The total population is split into a susceptible part $S = S(t, \boldsymbol{x})$ and an infected part $I = I(t, \boldsymbol{x})$:

$$P = S + I \ .$$

The disease transmission is assumed to be frequency-dependent; it occurs via proportionate mixing transmission (Nold, 1980; Hethcote, 2000; McCallum et al., 2001), i.e., the number of contacts between infected and susceptible individuals is constant, as it has been argued for rural/suburban cat populations (Fromont et al., 1998). The transmission coefficient is $\sigma > 0$ (per year). In contrast to HIV, the disease is assumed not to be transmitted to offspring. Hence, newborns of the infected are in the susceptible class. The infected suffer an additional disease-related mortality $\alpha > 0$ (per year), which shall be referred to as virulence. There is no recovery from the disease. Removal of infected is only by death. The corresponding transfer diagram is

$$\downarrow \beta(P)\,(S{+}I)$$
$$S \xrightarrow{\ \sigma \frac{SI}{S+I}\ } I$$
$$\downarrow \mu(P)\,S \qquad \downarrow [\mu(P){+}\alpha]\,I$$

The spatial propagation of the individuals is modelled by diffusion with diffusion coefficients $D_S \geq 0$ and $D_I \geq 0$ (km^2 per year) for the susceptibles and infected, respectively. This basic model is described by a system of two partial differential equations:

$$\frac{\partial S}{\partial t} = -\sigma\frac{SI}{P} + \beta(P)P - \mu(P)S + D_S\Delta S \,, \qquad (2.1)$$

$$\frac{\partial I}{\partial t} = +\sigma\frac{SI}{P} - \alpha I - \mu(P)I + D_I\Delta I \,, \qquad (2.2)$$

where Δ is the Laplacian. The initial and boundary conditions will be specified later.

System (2.1-2.2) is an extension of the model by Courchamp et al. (1995), which describes the transmission of FIV for the special case of a spatially homogeneous population with $D_S = D_I = 0$ and with logistic growth of the host population, i.e.,

$$\beta(P) = b > 0 \,, \qquad (2.3)$$

$$\mu(P) = m + \frac{r}{K}P \,, \quad r = b - m \,, \quad m > 0 \,, \qquad (2.4)$$

which yields the well-known linearly decreasing per-capita growth rate

$$g(P) = r(1 - P/K) \,.$$

Gao and Hethcote (1992) studied SIRS models with density-dependent birth and death rates. The above FIV model is a special case, for which they provided global stability results. Zhou and Hethcote (1994) moreover obtained global stability results for a model with generalized logistic growth. If

$$\beta(P) \text{ is nonincreasing,} \qquad (2.5)$$

$$\mu(P) \text{ is nondecreasing and} \qquad (2.6)$$

$$g(P) = 0 \text{ has a unique positive solution } K, \qquad (2.7)$$

then the per-capita growth rate $g(P)$ is nonincreasing, and K (the carrying capacity) is the unique stationary state of the disease-free, local model, which is globally asymptotically stable for $P(0) > 0$. The SI model with both logistic and generalized logistic growth has four stationary states and exhibits three different dynamics: eradication of the disease, emergence of a stable endemic stationary state or extinction of the host population. Periodic solutions do not exist, which has been proven with the Dulac criterion.

Next, a generalized strong Allee effect in the vital dynamics is considered, which is based upon the assumptions

$$\frac{d}{dP}\,g(P) \begin{cases} > 0 & \text{for } 0 < P < P_{opt} \,, \\ < 0 & \text{for } P_{opt} < P \,, \end{cases} \qquad (2.8)$$

$$g(P) = 0 \text{ has two positive roots } K_- \text{ and } K_+, \ 0 < K_- < P_{opt} < K_+ \,. \quad (2.9)$$

This yields a per-capita growth rate that is maximal at some intermediate density P_{opt}. The disease-free, local model has three nonnegative stationary states: K_- is unstable, K_+ is globally stable in the range $P(0) > P_{opt}$ while 0 is stable in the range $0 < P(0) < P_{opt}$. Furthermore, the following assumptions are posed:

$$\beta(P) \text{ is concave in } (0, K_+) \,, \qquad (2.10)$$

$$\mu(P) \text{ is nondecreasing and convex in } (0, K_+) \,, \qquad (2.11)$$

which imply that $g(P)$ is concave.

As a motivating example of the strong Allee effect, the following per-capita growth rate is considered:

$$g(P) = a(K_+ - P)(P - K_-) \, ,$$

which is quadratic with $P_{opt} = (K_+ + K_-)/2$. The parameter $a > 0$ scales the maximum per-capita growth rate. The fertility and mortality functions take the parametric forms

$$\beta(P) = \begin{cases} a(-P^2 + [K_+ + K_- + e]P + c) \, , & \text{for } 0 \le P \le K_+ + K_- \, , \\ \text{nonnegative and nonincreasing,} & \text{otherwise} \, , \end{cases} \quad (2.12)$$

$$\mu(P) = a(eP + K_+ K_- + c) \, . \tag{2.13}$$

The case differentiation in (2.12) is necessary, because the first case becomes negative for $P > K_+ + K_- + e$. We want to emphasize, that the first case is sufficient for the model analysis provided that the initial conditions are appropriate, e.g. $P \le K_+$. For biological reasons, however, we require for $P > K_+ + K_-$ that the fertility function remains nonnegative, nonincreasing as well as continuous. The mortality function linearly increases with P analogously to the logistic behaviour. The fertility function differs in being quadratic. It increases with P for small population densities, reaches a maximum at some intermediate population density and decreases with P for large population densities. The parameters $e, c \ge 0$ determine the effect of density-dependence and -independence, respectively. They do not affect, however, the intrinsic per-capita growth rate $g(P)$.

2.2.1 Parameter values

The parameter values are chosen as follows. The virulence $\alpha = 0.2$ year^{-1} is kept constant as in Courchamp et al. (1995). In the same study, the transmission coefficient has been estimated to be approximately $\sigma = 3.0$ year^{-1}. FIV is mainly transmitted through bites during aggressive contacts between cats. Hence, σ is related to the number of contacts between susceptibles and infected resulting in bites and eventually in virus transmission. Since there might be a large variation in the transmission coefficient, we will use σ in the following as bifurcation parameter. The parameters describing the logistic growth are $b = 2.4$ year^{-1}, $m = 0.6$ year^{-1} and $K = 200$ individuals per km^2 for rural cat populations, cf. Courchamp et al. (1995). The resulting fertility and mortality functions are shown in Fig. 2.1a. In the Allee effect model, parameterization is more difficult. First of all, $K_+ = 200$ individuals per km^2 is set to the same carrying capacity. Next, let us assume that there is an Allee threshold of ten percent, i.e., $K_- = 20$ individuals per km^2. The remaining parameters are chosen to yield (i) a similar natural mortality function as in the logistic model and (ii) a reduction in the fertility function with decreasing density, cf. Fig. 2.1b. The population growth rate exhibiting a strong Allee effect is plotted in Fig. 2.1c. Lastly, the diffusion coefficient has to be estimated. The relevant mechanism for population spread is considered to be dispersal (Fromont et al., 2003), since male individuals move under the pressure of dominant males from their native home range to vacant areas. Home range sizes vary from approximately 1 ha for urban stray cats to several hundred ha for populations in non-anthropized habitats, e.g. Fromont

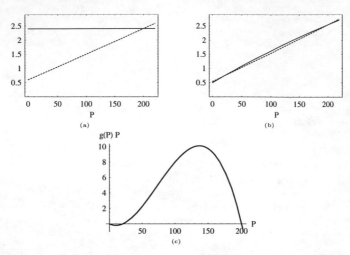

Figure 2.1: Fertility function $\beta(P)$ (solid lines) and mortality function $\mu(p)$ (dashed lines) of the models with (a) logistic growth and (b) a strong Allee effect. (c) Population growth rate $g(P) = \beta(P) - \mu(P)$ exhibiting a strong Allee effect. Parameters: $K = K_+ = 200, b = 2.4, m = 0.6, K_- = 20, a = 10^{-5}, c = 5 \cdot 10^4, e = 10^3$.

et al. (1998). Let us assume an observed mean net displacement of 1.77 km for populations of feral domestic cats in rural/suburban areas – having in mind a mean home range of 10 ha and that a dispersing individual has to pass on average five other home ranges before finding a vacant one. Then we can estimate a diffusivity of $D = 1$ km^2 per year, cf. eqn. (3.6) in Shigesada and Kawasaki (1997). Naturally, this is a very rough approximation, but the main aim is to explore possible patterns of spatiotemporal dynamics. Though we have the epizootiological problem of FIV spread in mind, we will keep throughout this paper the standard epidemiological notation such as *endemic* instead of *enzootic*, because the investigated model is rather general.

2.3   Stability results of the non-spatial model with Allee effect

In this section, the existence of the stationary states as well as their stability are summarized for the spatially homogeneous system with assumptions (2.8-2.11) for a generalized strong Allee effect. Details of the calculations can be found in the Appendix. System (2.1-2.2) with $D_S = D_I = 0$ exhibits a singularity in the transmission term. Introducing the prevalence $i = I/P \in [0,1]$ and

Table 2.1: Results of the stability analysis of the non-spatial system (2.14-2.15) with generalized Allee effect (2.8-2.11). The left column contains the stationary states (P^*, i^*). The other columns are divided according to the parameter regions along the ray for $\sigma - \alpha$ which are separated by the values in the top row. (*) The most right column corresponds to the case that additionally the function $\phi(P)$ defined in (2.22) does not achieve a positive maximum in (K_-, K_+). "l.a.s." stands for locally asymptotically stable, "g.a.s." for globally asymptotically stable, and "–" means that the stationary state does not exist or is not feasible.

	$\beta(0)$	$\beta(K_-)$		$\beta(K_+)$	(*)	$\sigma - \alpha$
$(0,0)$	l.a.s.	unstable	unstable	unstable	unstable	
$(0,i_2)$	–	l.a.s.	l.a.s.	l.a.s.	g.a.s.	
$(K_-,0)$	unstable	unstable	unstable	unstable	unstable	
$(K_+,0)$	l.a.s.	l.a.s.	l.a.s.	unstable	unstable	
(P_{3-},i_{3-})	–	–	unstable	unstable	–	
(P_{3+},i_{3+})	–	–	–	l.a.s.	–	
		extinction disease-free		extinction endemic	extinction	

reformulation in (P,i) state variables simplify matters:

$$\frac{\mathrm{d}P}{\mathrm{d}t} = [g(P) - \alpha\, i\,]\, P\,, \tag{2.14}$$

$$\frac{\mathrm{d}i}{\mathrm{d}t} = [\sigma - \alpha - \beta(P) - (\sigma - \alpha)\, i\,]\, i\,. \tag{2.15}$$

There are six stationary states. They are summarized along with the stability results in Table 2.1. For the sake of comparison, the stability results of the generalized logistic model are given in Table 2.2 as well. The formulation in (P,i) state variables allows to distinguish between the trivial extinction state $(0,0)$ and the disease-induced extinction state $(0,i_2)$ with an ultimate prevalence $i_2 > 0$. The latter one reflects that in the limit process $P \to 0$ there is a non-zero ultimate prevalence, and that the host population goes extinct as a consequence of infection with the disease (de Castro and Bolker, 2005). The stationary state $(K_+, 0)$ corresponds to the eradication of the disease, so that the disease-free population can settle down at its own carrying capacity. There are two equilibria, which do not exist in the logistic model. First, $(K_-, 0)$, which corresponds to the minimum viable population density in the disease-free Allee model, is always unstable. Second, there is an additional nontrivial stationary state. Denoting the larger and the smaller total population density with P_{3+} and P_{3-}, respectively, there are the nontrivial states (P_{3+}, i_{3+}) and (P_{3-}, i_{3-}).

One of the main results is that, again, sustained periodic oscillations are not possible in the generalized Allee effect model with proportionate mixing transmission. This follows from index theory and the fact that the only unstable interior equilibrium (P_{3-}, i_{3-}), around which a limit cycle could exist, is always saddle, cf. the Appendix.

It should be noted, that as soon as the initial total population verifies $P(0) < K_-$, it follows from (2.8-2.9) that the population goes extinct, i.e., $P(t) \to 0$ as $t \to \infty$; this dynamics is driven by the Allee effect alone and independent

Table 2.2: Results of the stability analysis of the non-spatial system (2.14-2.15) with generalized logistic behaviour (2.5-2.7).

$$\beta(0) \qquad \beta(K) \qquad \frac{\alpha\beta(0)}{\alpha-g(0)}$$

					$\sigma - \alpha$
$(0,0)$	unstable	unstable	unstable	unstable	
$(0,i_2)$	–	–	unstable	g.a.s.	
$(K,0)$	g.a.s.	unstable	unstable	unstable	
(P_3,i_3)	–	g.a.s.	g.a.s.	–	
	disease-free	endemic		extinction	

of the epidemics. Hence, there is always a stable extinction state with a basin of attraction containing at least $0 \leq P < K_-$ in the (P,i) phase plane. The extinction state has an ultimate prevalence of either zero or $i_2 > 0$. Note that the trivial state $(0,0)$ in the logistic model was always unstable.

Increasing the disease-related parameter combination $\sigma - \alpha$, various dynamical regimes can be observed. Due to the Allee effect, there are two bistable regimes, in which the asymptotic behaviour depends on the intial condition, and one monostable regime. These regimes are summarized in the bottom row of Table 2.1. They are: (i) population extinction or eradication of the disease, (ii) population extinction or endemicity of the disease, and (iii) disease-induced extinction. The last regime is also observed in the logistic case, cf. Table 2.2.

A closer look at Table 2.1 reveals that when $\sigma - \alpha < \beta(0)$, there are two locally stable stationary states $(0,0)$ and $(K_+,0)$. For $\beta(0) < \sigma-\alpha$, the disease-induced extinction state $(0,i_2)$ emerges and exchanges stability with the trivial solution. The unstable (saddle node) endemic state (P_{3-},i_{3-}) also exists when $\beta(K_-) < \sigma - \alpha < \beta(K_+)$. For $\sigma - \alpha > \beta(K_+)$, $(0,i_2)$ is still a locally stable stationary state. $(K_+,0)$ and $(K_-,0)$ are unstable. Two endemic states can exist together, the unstable one and a stable one (P_{3+},i_{3+}), wherein P_{3+} is the largest root of the function $\phi(P)$ defined in (2.22) within the range (K_-,K_+). Whenever this root exists, the prevalence is given by $i_{3+} = g(P_{3+})/\alpha$, see (2.21). When $\phi(P)$ does not achieve a positive maximum in the range (K_-,K_+), there are no nontrivial states, and the disease-induced extinction state is globally stable.

It is instructive to consider the nullclines in the phase plane. In Fig. 2.2, this is done for the motivating example (2.12-2.13). The trivial nullclines are on the axes, while the nontrivial nullclines are quadratics, one open to the bottom and one open to the top. They intersect either in a non-feasible region (a), in a single nontrivial equilibrium (b), in two nontrivial equilibria (c) or in none nontrivial equilibrium (d). Though there is no stationary state in the latter case, the temporal dynamics exhibit a slow-down of the trajectories in the region where the nullclines are close to each other. This results in a "transient" epidemic, which is shown in Fig. 2.3

As noted above, when $0 < P(0) < K_-$, the population dies out by the Allee effect alone. But also for $P(0) > K_-$ numerical simulations show that one may have extinction of the population for a suitable choice of $(P(0),i(0))$. This is a consequence of the joint interplay between the population reduction due to the disease and the Allee effect vital dynamics. While in the model with logistic growth extinction is only possible if $\sigma - \alpha$ is large, the Allee effect generally

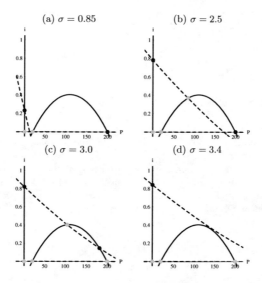

Figure 2.2: Nullclines and stationary states in model (2.14-2.15) with Allee effect (2.12-2.13). The solid and the dashed lines are the nullclines of P and i, respectively. Black points are stable equilibria and grey points are unstable equilibria. Other parameter values as in Fig. 2.1 and $\alpha = 0.2$. Cases (a-d) correspond to the four columns from the right-hand side in Table 2.1.

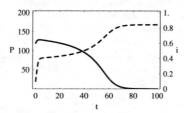

Figure 2.3: Transient epidemic in model (2.14-2.15) with Allee effect (2.12-2.13), when the nullclines do not intersect and the disease-induced extinction state is globally stable. The solid and the dashed lines correspond to the course of P and i, respectively. Parameters as in Fig. 2.2d, $P(0) = 120, i(0) = 0.1$.

makes possible the population extinction. Hence, the Allee effect is especially important in the parameter ranges, where the logistic model allows endemicity. In turn, when the dynamics is mainly driven by the losses due to the disease, i.e., $(0, i_2)$ is globally stable, the asymptotic behaviour of the logistic and the Allee model is similar.

2.4   Spatial spread

Additionally accounting for diffusion as spatial propagation mechanism, this section takes into account the full system (2.1-2.2). Infected individuals are assumed not to be affected by the disease in their mobility, thus $D = D_S = D_I$. Moreover, considerations are restricted to the one-dimensional space, i.e., the Laplacian is set to $\Delta = \partial^2/\partial x^2$. Throughout this section, no-flux boundary conditions are assumed. The initial conditions generally distinguish between a left, middle and right region (which shall later reflect the wake of the host population invasion front in which the disease is introduced, the disease-free invasion front and empty space, respectively):

$$(S(0, x), I(0, x)) = \begin{cases} (S_l, I_l) & \text{if } x < x_l \,, \\ (S_l, 0) & \text{if } x_l \leq x < x_r \,, \quad 0 < x_l \leq x_r \,. \\ (S_r, 0) & \text{if } x \geq x_r \,, \end{cases} \tag{2.16}$$

Throughout this paper, S_l will be set to the carrying capacity, i.e. $S_l = K$ for the model with logistic growth and $S_l = K_+$ for the model with Allee effect, and $I_l = 1$. When the host population invades empty space, $S_r = 0$, and if the diseases is introduced into a completely established population, $S_r = S_l$.

For the numerical simulations, the Runge-Kutta scheme of fourth order is applied for the reaction part and an explicit Euler scheme for the diffusion part of the PDE. In order to handle the singularity at the extinction state, a small threshold $\delta = 10^{-10}$ is applied. If $S(t) + I(t) < \delta$, then the transmission terms in (2.1-2.2) are neglected. In order to suppress effects resulting from a microscopically small band of individuals propagating ahead the actual fronts ("atto-fox problem", (Mollison, 1991)), an additional threshold $\epsilon = 10^{-5}$ is applied, below which the population densities of both susceptibles and individuals are simply reset to zero. The value of this threshold has been chosen to weaken atto-effects, but not to change the qualitative behaviour of the waves.

First it should be noted that in the disease-free model ($I = 0$) travelling frontal waves with constant speed and shape emerge (Fisher, 1937; Kolmogorov et al., 1937; Aronson and Weinberger, 1975; Lewis and Kareiva, 1993). There is a minimum wave speed in the model with logistic growth and a unique wave speed in the model with Allee effect, which respectively are:

$$v = 2\sqrt{rD} \,, \tag{2.17}$$

$$v = \sqrt{2aD} \, (K_+/2 - K_-) \,. \tag{2.18}$$

There are two particularities in the Allee effect model.

1. When the Allee threshold $K_- > K_+/2$, then the population wave moves backward.

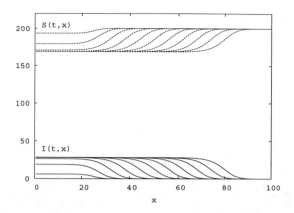

Figure 2.4: Travelling infection wave in the SI model with logistic growth (2.3-2.4). The solid lines represent I, the dotted lines S. Displayed are different snapshots, which have been taken in a time interval with a delay of 5, beginning at $t = 5$ and ending at $t = 55$ (from left to right). Parameters as in Fig. 2.1a, $\sigma = 3.0, \alpha = 0.2, S_r = K, x_l = x_r = 5$.

2. Such one-component bistable reaction-diffusion systems are known to show nucleation-type behaviour, i.e., nucleii of a stable "phase" $P = K_+$ in an unstable phase $P = 0$ will decay unless they have reached a certain critical size, cf. (Nitzan et al., 1974; Malchow and Schimansky-Geier, 1985; van Saarlos, 2003). Otherwise, they advance with the asymptotic rate of spread given in (2.18).

2.4.1 Spread in a settled disease-free population

The situation is considered that the disease-free population has established in all the space at carrying capacity, i.e. $S_r = K$ or $S_r = K_+$, respectively, and that infected individuals are at the left-hand boundary as specified in (2.16).

The numerical simulation of the model with logistic growth is shown in Fig. 2.4. A travelling infection wave emerges and advances with a constant speed v. In its wake, the population settles down to the endemic state. The infection front propagates with a speed of $v = 1.2$ km/year. This matches well the wave speed which can also be easily derived by linearization of the equations "at the leading edge", i.e., far in front of the travelling front where $S \approx K$ and $I \approx 0$. Then the equation for I in (2.1-2.2) reduces to be of Skellam/Luther type (Skellam, 1951; Luther, 1906), and the minimum wave speed is

$$v = 2\sqrt{(\sigma - \alpha - b)\, D}\ . \tag{2.19}$$

In the model with Allee effect, the emergence and propagation of a travelling wave can be observed, too. With the same approach as above, one obtains the wave speed

$$v = 2\sqrt{(\sigma - \alpha - a[c + (e + K_-)K_+])\, D}\ . \tag{2.20}$$

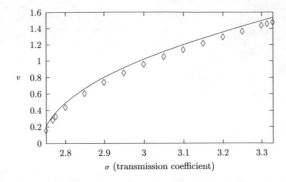

Figure 2.5: Wave speed of disease spread in an established host population with Allee effect. The line is the approximation (2.20) and the data points are numerical results. Parameter values as in Fig. 2.1 with $\alpha = 0.2$. The values for the transmission coefficient σ have been chosen according to the existence of the endemic state (P_{3+}, i_{3+}).

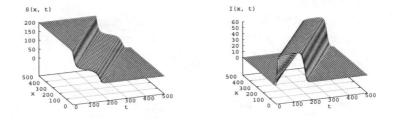

Figure 2.6: Travelling fatal epidemic with succeeding host population extinction. Model with Allee effect, parameter values as in Fig. 2.2d, $S_r = K_+, x_l = x_r = 10$.

This is an approximation and has to be taken with caution, because the travelling front in the Allee effect model is a "pushed" wave. Because the infection spreads within a population which has established at carrying capacity, we argue that (2.20) may be a reasonable approximation provided that the coexistence state is large enough. In order to check the robustness of (2.20) against P_{3+}, numerical simulations were run with varying transmission coefficient σ. The results in Fig. 2.5 show a very good accordance. However, we want to emphasize that in other parameter ranges where P_{3+} is closer to the Allee threshold regions the match possibly might not be as good.

Both wave speed approximations include the disease-related parameters and the diffusivity. In the model with logistic growth, only the birth rate is additionally included. This reflects that the wave is "pulled", whereas the Allee effect-wave is "pushed", and therefore also the other vital parameters play a role, e.g., the Allee threshold and the carrying capacity.

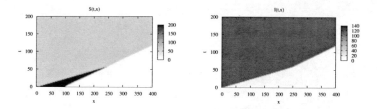

Figure 2.7: Travelling infection wave introduced in the wake of an invading host population with logistic growth. When the disease catches up the invading host population front, the disease spread is slowed down, but the host population (now endemic) still advances with the same speed. Parameters as in Fig. 2.4, but with $S_r = 0, x_l = 5, x_r = 100$ and $\sigma = 7$, in order to enable the catch-up.

Next, if the transmission coefficient is further increased than in Fig. 2.5, the nontrivial states disappear and the disease-induced extinction state $(0, i_2)$ is globally stable, cf. Table 2.1. Numerical simulations show the propagation of travelling pulse-like epidemics that wipe out the host population as displayed in Fig. 2.6. This effect is a result of the "transient" epidemic, cf. Fig. 2.3. The length of the pulse depends on how closely the nontrivial nullclines approach each other (Fig. 2.2d). The travelling epidemic with succeeding population extinction is associated with the stability of the disease-induced extinction state. Hence, this fatal epidemic wave can also be obtained in the model with logistic growth if the parameters are appropriately chosen, cf. Table 2.2.

2.4.2 Spread in colonizing population

Now, the host population is assumed still to be in a colonizing process, i.e. $S_r = 0$. The disease is subsequently introduced in the wake of the host population front, cf. (2.16).

In the logistic SI model with the FIV parameters, the speed (2.17) of the invading host population is larger than the speed (2.19) of the disease. Thus, the distance between the two fronts increases with time. However, the infection wave can catch up the disease-free front, if $\sigma - \alpha > 2b - m$. This is illustrated in Fig. 2.7. For a better visualization, the waves are now displayed in the (x, t) plane with a grey colour shading according to the densities of susceptibles and infected. The infection and the host population invasion front combine to a travelling front of the endemic state into empty space. The numerical results show that this front moves with the same speed (2.17) as the invading host population. This means that the infection spread rate is reduced to this speed, because of which there is a kink in the expansion of I in Fig. 2.7.

In the model with Allee effect, three different types of spatiotemporal dynamics can be observed when a catch-up has taken place. First, the endemic front does not move with the same speed (2.18) as the disease-free invasion front before. Instead, the endemic front either slows down or becomes recessive. The latter case of front reversal is shown in Fig. 2.8. Shortly before the catch-up, the remaining "atto-individuals" cause a hump of infection ahead of the actual

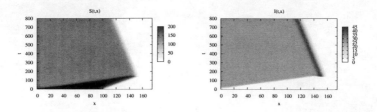

Figure 2.8: Front reversal in the model with Allee effect, when the infected front catches up the host population front. Parameters as in Fig. 2.2c with $S_r = 0, x_l = 5, x_r = 100$.

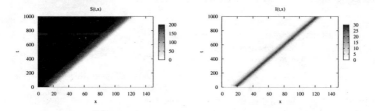

Figure 2.9: Infection pulse in the model with Allee effect. Parameters as in Fig. 2.8, except $\sigma = 2.5$ and $x_l = 10, x_r = 15$.

infection front. Then, there is a retreat of the total population, which corresponds to a wave of extinction, though the coexistence state is locally stable. Note the increased density of infected at the head of the retreating front. A detailed study of this phenomenon is given in Hilker et al. (2005).

Second, when there is a unique, unstable nontrivial equilibrium as in Fig. 2.2b, a travelling infection pulse emerges, which propagates jointly with the front of the host population. This corresponds to a travelling epidemic and is illustrated in Fig. 2.9. Though the infection pulse continues its advancement as long as there are no boundary restrictions, the disease fades out at a fixed location in space when the pulse has passed. Then, the population approaches the carrying capacity again – in contrast to the fatal epidemic with succeeding population extinction shown in Fig. 2.6. It should be noted, that the emergence of this travelling pulse requires appropriate values of σ, initial conditions as well as x_l and x_r being close enough to each other.

Third, consider the parameter region where the disease-induced extinction state is globally stable and a transient epidemic develops as in Fig. 2.3. Then the disease-free population spreads in an invasion front, and the introduced disease causes a travelling epidemic before the ultimate extinction due to the transient dynamics, cf. Fig. 2.10. Thus far this effect is the same as the fatal epidemic (Fig. 2.6). The finite initial distribution of the host population, however, induces the disappearance of the epidemic wave. When the infection catches up the invading host population, there is again a front reversal of the endemic state,

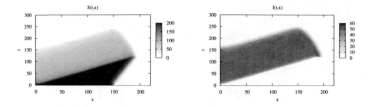

Figure 2.10: Spatially restricted epidemic in the model with Allee effect. In the non-spatial model, there is no non-trivial state and the disease-induced extinction state is globally stable. Parameters as in Fig. 2.8, except $\sigma = 3.4$, what yields the local dynamics as in Fig. 2.3. $x_l = 5, x_r = 150$.

and thus a travelling wave of extinction limits the propagation of the epidemics. This results in the spatial extinction of the total population, and the epidemic could only spread within a spatially restricted area. Note that this effect is also possible in the model with logistic growth since it is associated with the disease-induced extinction state as well as appropriate initial conditions.

2.5   Discussion and conclusions

This paper has investigated the impact of a strong Allee effect in the vital dynamics of an epidemiological SI model upon the temporal and spatiotemporal disease spread as well as on the host population. First of all, the Allee effect induces a threshold value in the initial conditions below which the population dies out (minimum viable population size) and a critical value of the spatial length of an initial nucleus (also called the problem of critical aggregation, cf. Petrovskii and Shigesada (2001)). Numerical simulations show that both phenomena are strengthened in the presence of the disease.

Furthermore, the Allee effect leads to bistability in the local transmission dynamics. Jointly with the minimum viable population size, this has severe implications for possible control methods, since they do not necessarily rely on reducing the basic reproductive ratio anymore. Instead, disease control could be established by shifting the system's trajectory in the desired domain of attraction, which may be easier to manage than modifying parameter regimes.

If the infectiousness of the disease, i.e., a high transmissibility and/or a small enough virulence, is large in comparison with the demographic reproductiveness, the Allee effect becomes less important, because the population dynamics is dominantly driven by the disease to extinction. This disease-induced extinction state is typical for epidemiological models with proportionate mixing transmission and has also been found in models with vital dynamics of logistic, exponential or recruitment type. Similarly, sustained oscillations could not be found in the model with Allee effect, either. There are, however, three other additional features: (i) The trivial state can be locally stable. (ii) There is a second nontrivial state, which is always unstable. (iii) In the parameter range of disease-induced extinction, there is still an epidemic possible, if the nontrivial

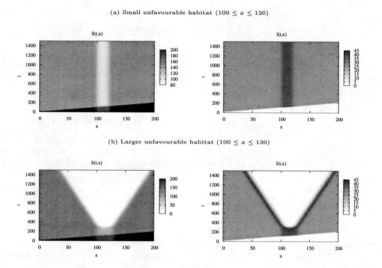

Figure 2.11: (a) Diffusion-enhanced survival and (b) population extinction emanating from the unfavourable habitat in the spatially heterogeneous model with Allee effect. Parameters as in Fig. 2.2c, i.e., the endemic state (P_{3+}, i_{3+}) is stable with $K_+(x) = 200$. In the unfavourable habitats: $K_+(x) = 185$, i.e., the disease-induced extinction state is globally stable (similarly to Fig. 2.2d). $S_r = K_+, x_l = x_r = 5$.

nullclines are close to each other. The importance of transient dynamics has recently been highlighted in Hastings (2004).

The emergence and propagation of travelling frontal waves of infection have been numerically observed in a spatial setting. Analytical wave speed approximations have been derived for the cases when the disease propagates in an established host population. When the host population itself still colonizes empty space, various spatiotemporal dynamics can be observed. In the model with logistic growth the infection front is slowed down to the constant invasion speed of the host population, which is not affected by this catch-up. In contrast, in the model with Allee effect, the propagation of the host population front is either slowed down or reversed. Thus, the extinction of the invasive host population is possible when the disease is subsequently introduced. This is a mechanism which can be attributed to the joint interplay of the Allee effect, the disease and the spatial diffusion. Thus, the virus might be a potential biocontrol agent, cf. Fagan et al. (2002). A similar dynamics has been observed in predator-prey models in which the prey exhibits an Allee effect (Owen and Lewis, 2001; Petrovskii et al., 2005b). An analytical investigation of this effect is challenging because of the set of two PDEs with cubic nonlinearity, but possible approaches are presented in Hilker et al. (2005). Pathogen-driven host extinction in a spatial context has also been observed in a model of lattice structured populations (Sato et al., 1994; Haraguchi and Sasaki, 2000).

Moreover, travelling infection pulses in front of the colonizing host population can be observed if the unique endemic state is unstable in the local dynamics. This is an additional feature of the model with Allee effect. Both models with Allee effect and logistic growth can exhibit travelling fatal epidemics with host extinction in the parameter regime with the "transient" epidemic. This is associated with the globally stable disease-induced extinction state. If the intial distribution of the host population is finite, a spatially restricted infection front appears before ultimate population extinction.

Future work will also have to consider spatial heterogeneity. By way of example, we briefly address the specific case of singular spatial inhomogeneities in the parameter distribution, in order to obtain a rough impression about possible effects. Therefore, a less favourable/hostile habitat patch is assumed in which the population cannot persist due to local conditions. These considerations are restricted to the model with Allee effect, because in the logistic case no novel effects could be observed. The carrying capacity $K_+ = K_+(x)$ is now space-dependent and reduced in the unfavourable habitat, cf. Fig. 2.11. At the beginning, all habitats are inhabited by the disease-free population at carrying capacity, and the disease is introduced at the left-hand side, i.e. $S_r = K_+$ in (2.16). If the length of the habitat is below a critical value (Fig. 2.11a), then the diffusion of individuals from neighbouring locations enhances the survival at a level of high infection in the unfavourable habitat, in which the population would locally go extinct. Otherwise, cf. Fig. 2.11b, the population dies out in the unfavourable habitat after a transient epidemic phase (cf. Fig. 2.3). Then, a travelling wave of extinction, similarly to the front reversal scenario, emerges and propagates in both directions until global population extinction. These extinction waves are of considerable interest, since they destabilize coexistence states which are locally stable in local dynamics. Here, two important conclusions are to be drawn: (i) They can be induced by spatial heterogeneity, and (ii) there seems to be a critical spatial length of the heterogeneity.

Overall, the Allee effect induces a rich variety of (spatio-)temporal dynamics in the considered epidemiological model. Since it can have significant consequences on the fate of epidemics, endemics and invasions, its role has to be further investigated. Of particular interest would be the robustness of the current results against other epidemiological details such as other transmission functions, vertical transmission or disease-related reduced fertility. The extension of an exposed compartment, for instance, enables sustained oscillations if disease transmission is of mass action type (Anderson et al., 1981). Recent results indicate that the Allee effect makes possible limit cycle oscillations in an SI model with mass action transmission (Hilker et al., in prep.), even if there is no exposed compartment.

Acknowledgements

The authors thank two anonymous reviewers for their comments which helped improving the paper. This work was initiated in February 1996 while M.L. was holding a visiting position at Kyushu University under a grant from the Japan Society for the Promotions of Science (JSPS). S.P. acknowledges partial support from the University of California Agricultural Experiment Station through Professor Bai-Lian Li.

References

Abramson, G., Kenkre, V. M., Yates, T. L., Parmenter, R. R. (2003). Traveling waves of infection in the hantavirus epidemics. *Bulletin of Mathematical Biology* 65, 519–534.

Anderson, R. M., Jackson, H. C., May, R. M., Smith, A. M. (1981). Population dynamics of foxes rabies in europe. *Nature* 289, 765–770.

Anderson, R. M., May, R. M. (1979). Population biology of infectious diseases: Part I. *Nature* 280, 361–367.

Aronson, D. G., Weinberger, H. F. (1975). Nonlinear diffusion in population genetics, combustion, and nerve propagation. In *Partial differential equations and related topics* (Goldstein, J. A., ed.). No. 446 in Lecture Notes in Mathematics, Springer-Verlag, Berlin, pp. 5–49.

Brauer, F. (1990). Models for the spread of universally fatal diseases. *Journal of Mathematical Biology* 28, 451–462.

Brauer, F., Castillo-Chavez, C. (2001). *Mathematical models in population biology and epidemiology*. Springer-Verlag, New York.

Busenberg, S., van den Driessche, P. (1990). Analysis of a disease transmission model in a population with varying size. *Journal of Mathematical Biology* 28, 257–270.

Cantrell, R. S., Cosner, C. (2003). *Spatial ecology via reaction-diffusion equations*. Wiley, Chichester.

Caraco, T., Glavanakov, S., Chen, G., Flaherty, J. E., Ohsumi, T. K., Szymanski, B. K. (2002). Stage-structured infection transmission and a spatial epidemic: a model for lyme disease. *The American Naturalist* 160(3), 348–359.

Courchamp, F., Clutton-Brock, T., Grenfell, B. (1999). Inverse density dependence and the Allee effect. *Trends in Ecology & Evolution* 14(10), 405–410.

Courchamp, F., Cornell, S. J. (2000). Virus-vectored immunocontraception to control feral cats on islands: a mathematical model. *Journal of Applied Ecology* 37, 903–913.

Courchamp, F., Pontier, D., Langlais, M., Artois, M. (1995). Population dynamics of Feline Immunodeficiency Virus within cat populations. *Journal of Theoretical Biology* 175(4), 553–560.

Courchamp, F., Sugihara, G. (1999). Modeling the biological control of an alien predator to protect island species from extinction. *Ecological Applications* 9(1), 112–123.

de Castro, F., Bolker, B. (2005). Mechanisms of disease-induced extinction. *Ecology Letters* 8, 117–126.

Dennis, B. (1989). Allee effects: population growth, critical density, and the chance of extinction. *Natural Resource Modeling* 3, 481–538.

Diekmann, O., Heesterbeek, J. A. P. (2000). *Mathematical epidemiology of infectious diseases. Model building, analysis and interpretation.* John Wiley & Son, New York.

Diekmann, O., Kretzschmar, M. (1991). Patterns in the effects of infectious diseases on population growth. *Journal of Mathematical Biology* 29, 539–570.

Dwyer, G. (1994). Density dependence and spatial structure in the dynamics of insect pathogens. *The American Naturalist* 143(4), 533–562.

Fagan, W. F., Lewis, M. A., Neubert, M. G., van den Driessche, P. (2002). Invasion theory and biological control. *Ecology Letters* 5, 148–158.

Fisher, R. A. (1937). The wave of advance of advantageous genes. *Annals of Eugenics* 7, 355–369.

Fromont, E., Pontier, D., Langlais, M. (1998). Dynamics of a feline retrovirus (FeLV) in host populations with variable spatial structure. *Proceedings of the Royal Society of London B* 265, 1097–1104.

Fromont, E., Pontier, D., Langlais, M. (2003). Disease propagation in connected host populations with density-dependent dynamics: the case of the Feline Leukemia Virus. *Journal of Theoretical Biology* 223, 465–475.

Gao, L. Q., Hethcote, H. W. (1992). Disease transmission models with density-dependent demographics. *Journal of Mathematical Biology* 30, 717–731.

Greenhalgh, D., Das, R. (1995). Modelling epidemics with variable contact rates. *Theoretical Population Biology* 47, 129–179.

Haraguchi, Y., Sasaki, A. (2000). The evolution of parasite virulence and transmission rate in a spatially structured population. *Journal of Theoretical Biology* 203, 85–96.

Hastings, A. (2004). Transients: the key to long-term ecological understanding? *Trends in Ecology & Evolution* 19, 39–45.

Hethcote, H. W. (2000). The mathematics of infectious diseases. *SIAM Review* 42(4), 599–653.

Hilker, F. M., Lewis, M. A., Seno, H., Langlais, M., Malchow, H. (2005). Pathogens can slow down or reverse invasion fronts of their hosts. *Biological Invasions* 7(5), 817–832.

Kermack, W. O., McKendrick, A. G. (1927). Contribution to the mathematical theory of epidemics, part I. *Proceedings of the Royal Society A* 115, 700–721.

Kermack, W. O., McKendrick, A. G. (1932). Contribution to the mathematical theory of epidemics. II - The problem of endemicity. *Proceedings of the Royal Society A* 138, 55–83.

Kolmogorov, A. N., Petrovskii, I. G., Piskunov, N. S. (1937). Étude de l'equation de la diffusion avec croissance de la quantité de matière et son application à un problème biologique. *Bulletin Université d'Etat à Moscou, Série internationale, section A* 1, 1–25.

Lewis, M. A., Kareiva, P. (1993). Allee dynamics and the spread of invading organisms. *Theoretical Population Biology* 43, 141–158.

Luther, R. (1906). Räumliche Ausbreitung chemischer Reaktionen. *Zeitschrift für Elektrochemie* 12, 596–600.

Malchow, H., Schimansky-Geier, L. (1985). *Noise and diffusion in bistable nonequilibrium systems*. No. 5 in Teubner-Texte zur Physik, Teubner-Verlag, Leipzig.

McCallum, H., Barlow, N., Hone, J. (2001). How should pathogen transmission be modelled? *Trends in Ecology & Evolution* 16(6), 295–300.

Mena-Lorca, J., Hethcote, H. W. (1992). Dynamic models of infectious diseases as regulator of population sizes. *Journal of Mathematical Biology* 30, 693–716.

Mollison, D. (1977). Spatial contact models for ecological and epidemics spread. *Journal of the Royal Statistical Society B* 39(3), 283–326.

Mollison, D. (1991). Dependence of epidemics and populations velocities on basic parameters. *Mathematical Biosciences* 107, 255–287.

Murray, J. D. (2002). *Mathematical biology. I: An introduction*. Springer-Verlag, Berlin, 3rd edition.

Murray, J. D. (2003). *Mathematical biology. II: Spatial models and biomedical applications*. Springer-Verlag, Berlin, 3rd edition.

Murray, J. D., Stanley, E. A., Brown, D. L. (1986). On the spatial spread of rabies among foxes. *Proceedings of the Royal Society of London B* 229, 111–150.

Nitzan, A., Ortoleva, P., Ross, J. (1974). Nucleation in systems with multiple stationary states. *Faraday Symposia of The Chemical Society* 9, 241–253.

Noble, J. V. (1974). Geographic and temporal development of plagues. *Nature* 250, 726–729.

Nold, A. (1980). Heterogeneity in disease-transmission modeling. *Mathematical Biosciences* 52, 227–2402.

Okubo, A., Maini, P. K., Williamson, M. H., Murray, J. D. (1989). On the spatial spread of the gray squirrel in Britain. *Proceedings of the Royal Society of London B* 238, 113–125.

Owen, M. R., Lewis, M. A. (2001). How predation can slow, stop or reverse a prey invasion. *Bulletin of Mathematical Biology* 63, 655–684.

Petrovskii, S., Shigesada, N. (2001). Some exact solutions of a generalized Fisher equation related to the problem of biological invasion. *Mathematical Biosciences* 172, 73–94.

Petrovskii, S. V., Malchow, H., Hilker, F. M., Venturino, E. (2005a). Patterns of patchy spread in deterministic and stochastic models of biological invasion and biological control. *Biological Invasions* 7, 771–793.

Petrovskii, S. V., Malchow, H., Li, B.-L. (2005b). An exact solution of a difffusive predator-prey system. *Proceedings of the Royal Society of London A* 461, 1029–1053.

Pugliese, A. (1990). Population models for diseases with no recovery. *Journal of Mathematical Biology* 28, 65–82.

Rass, L., Radcliff, J. (2003). *Spatial deterministic epidemics*. American Mathematical Society, Providence RI.

Sato, K., Matsuda, H., Sasaki, A. (1994). Pathogen invasion and host extinction in lattice structured populations. *Journal of Mathematical Biology* 32, 251–268.

Shigesada, N., Kawasaki, K. (1997). *Biological invasions: Theory and practice*. Oxford University Press, Oxford.

Skellam, J. G. (1951). Random dispersal in theoretical populations. *Biometrika* 38, 196–218.

Stephens, P. A., Sutherland, W. J. (1999). Consequences of the Allee effect for behaviour, ecology and conservation. *Trends in Ecology & Evolution* 14(10), 401–405.

Stephens, P. A., Sutherland, W. J., Freckleton, R. P. (1999). What is the Allee effect? *Oikos* 87(1), 185–190.

Thieme, H. R. (2003). *Mathematics in population biology*. Princeton University Press, Princeton NJ.

van den Bosch, F., Metz, J. A. J., Diekmann, O. (1990). The velocity of spatial population expansion. *Journal of Mathematical Biology* 28, 529–565.

van Saarlos, W. (2003). Front propagation into unstable states. *Physics Reports* 386, 29–222.

Wiggins, S. (2003). *Introduction to applied nonlinear dynamical systems and chaos*. Springer-Verlag, New York, 2nd edition.

Yachi, S., Kawasaki, K., Shigesada, N., Teramoto, E. (1989). Spatial patterns of propagating waves of fox rabies. *Forma* 4, 3–12.

Zhou, J., Hethcote, H. W. (1994). Population size dependent incidence in models for diseases without immunity. *Journal of Mathematical Biology* 32, 809–834.

Appendix: Stability analysis of the non-spatial model

We consider the ODE system (2.14-2.15) with the assumptions (2.8-2.11) for a generalized strong Allee effect. For the sake of simplicity, let the prime denote differentiation with respect to P. We will apply this notation to $g(P)$, $\beta(P)$ and $\phi(P)$.

One may check that the domain $P \geq 0$ and $0 \leq i \leq 1$ is invariant; moreover, no trajectory starting at $P(t=0) > 0$ and $i(t=0) > 0$ may hit the boundary $P = 0$ or the boundary $i = 0$ in finite time. There are four (semi-)trivial stationary states:

- $P_0 = 0$ and $i_0 = 0$,

- $P_{1+} = K_+$ and $i_{1+} = 0$,

- $P_{1-} = K_-$ and $i_{1-} = 0$,

- $P_2 = 0$ and $i_2 = [\sigma - \alpha - \beta(0)]/(\sigma - \alpha)$; it is feasible, i.e., $0 < i_2 \leq 1$, if and only if $\sigma - \alpha > \beta(0)$.

Looking for a nontrivial stationary state with $0 < P_3 \leq K_+$ and $0 < i_3 \leq 1$ one finds

$$i_3 = \frac{g(P_3)}{\alpha} = \frac{\sigma - \alpha - \beta(P_3)}{\sigma - \alpha} \; ; \tag{2.21}$$

Thus, a necessary condition to have a nontrivial stationary state is $\sigma - \alpha > 0$. Introducing the function $\phi(P)$ defined as

$$\phi(P) = (\sigma - \alpha)g(P) + \alpha\beta(P) - \alpha(\sigma - \alpha) \; ,$$

one is left with finding P_3 as a root of

$$\phi(P_3) = 0 \text{ in } (K_-, K_+) \; , \tag{2.22}$$

because, by (2.8-2.9), $g(P)$ is nonpositive in $(0, K_-)$. Still assuming $\sigma - \alpha > 0$, from conditions (2.10-2.11), $\phi(P)$ is concave in this range; hence there are either 0, 1 or 2 feasible roots. Let us denote P_{3-} the root located on the increasing branch of ϕ and P_{3+} the root located on the decreasing branch of ϕ, when such roots exist. In order to be a little bit more precise, let us recall that $g(0) < 0$ and $g(K_-) = g(K_+) = 0$; thus, using the monotonicity of the death-rate $\mu(P)$, it follows

$$0 \le \beta(0) < \beta(K_-) \le \beta(K_+) \text{ and } \phi(K_-) \le \phi(K_+) \ .$$

Hence, still assuming $\sigma - \alpha > 0$, one has

1. no nontrivial stationary solution when $\phi(K_-) \ge 0$, say $0 < \sigma - \alpha \le \beta(K_-)$;

2. a unique nontrivial stationary solution (P_{3-}, i_{3-}), when $\phi(K_-) < 0 \le \phi(K_+)$, say $\beta(K_-) < \sigma - \alpha \le \beta(K_+)$;

3. two nontrivial stationary solutions, labelled (P_{3-}, i_{3-}) and (P_{3+}, i_{3+}), or

4. none when $\phi(K_-) \le \phi(K_+) < 0$, say $\beta(K_-) \le \beta(K_+) < \sigma - \alpha$; this depends on whether ϕ achieves a positive maximum in (K_-, K_+) or not.

Figure 2.12: The number of nontrivial stationary states in model (2.14-2.15) with Allee effect (2.12-2.13) depends on the location of $\phi(P)$. Four different cases are possible. Their numbering refers to the description in the text.

This is illustrated in Fig. 2.12. It should be noted that $g'(P_{3+}) < 0$: assuming the opposite yields $\beta'(P_{3+}) > 0$ and therefore $\phi'(P_{3+}) > 0$, a contradiction.

The local stability analysis of the (semi-)trivial stationary states follows from the computation of the Jacobian matrix that is either a diagonal or a triangular matrix. The results are summarized in Table 2.1. Finally, at any nontrivial stationary state

$$J(P_3, i_3) = \begin{pmatrix} g'(P_3)P_3 & -\alpha P_3 \\ -\beta'(P_3)i_3 & -(\sigma - \alpha)i_3 \end{pmatrix} \ .$$

One has $\phi'(P_{3+}) < 0$, so that the determinant of $J(P_{3+}, i_{3+})$ is positive; its trace is negative because as noted above $g'(P_{3+}) < 0$ and $\sigma - \alpha > 0$ as soon as P_{3+} is feasible, yielding the local stability of (P_{3+}, i_{3+}) as soon as it is feasible. Conversely, $\det J(i_3, P_3) = -i_3 P_3 \phi'(P_3)$ so that (P_{3-}, i_{3-}) is unstable when it is feasible because $\phi'(P_{3-}) > 0$. To be more precise, (P_{3-}, i_{0-}) is always a saddle. Using this fact, we obtain from index theory that there cannot be any limit cycles. The index theory can give insight into the qualitative behaviour of closed orbits and multiple equilibria of planar dynamical systems (Brauer and Castillo-Chavez, 2001; Wiggins, 2003). In particular, the index of any simple closed curve is equal to the sum of the indices of all equilibria in the interior of the curve. The index of a periodic orbit is +1. Within the positive interior, there exist either only (P_{3-}, i_{3-}) that is always a saddle with index -1 or also (P_{3+}, i_{3+}) that is always stable with index +1. Hence, there cannot be any periodic orbit – neither around (P_{3-}, i_{3-}) nor around both (P_{3-}, i_{3-}) and (P_{3+}, i_{3+}). Note that a homoclinic orbit should not be treated as a periodic orbit for the application of index theory (Wiggins, 2003, p. 88), because of which we restrict or conclusion to the nonexistence of limit cycle oscillations.

Chapter 3

Pathogens can slow down or reverse invasion fronts of their hosts

Frank M. Hilker[1,2], Mark A. Lewis[3], Hiromi Seno[2], Michel Langlais[4] and Horst Malchow[1]

[1] Institute of Environmental Systems Research, Department of Mathematics and Computer Science, University of Osnabrück, 49069 Osnabrück, Germany

[2] Department of Mathematical and Life Sciences, Graduate School of Science, Hiroshima University, Kagamiyama 1-3-1, Higashi-hiroshima, 739-8526, Japan

[3] Department of Mathematical and Statistical Sciences, Department of Biological Sciences, University of Alberta, Edmonton T6G 2G1, Canada

[4] UMR CNRS 5466, Mathématiques Appliquées de Bordeaux, Université Victor Segalen Bordeaux 2, 33076 Bordeaux Cedex, France

Published in *Biological Invasions* 7(5), 817-832 (2005).[1]

Abstract

Infectious diseases are often regarded as possible explanations for the sudden collapse of biological invasions. This phenomenon is characterized by a host species, which firstly can successfully establish in a non-native habitat, but then spontaneously disappears again. This study proposes a reaction-diffusion model consisting of a simple SI disease with vital dynamics of Allee effect type. By way of travelling wave analysis, conditions are derived under which the invasion of the host population is slowed down, stopped or reversed as a consequence of a subsequently introduced disease. Hence, pathogens can drastically control the rate of spread of invasive species.

3.1   Introduction

Biological invasions are regarded as one of the most severe ecological problems, being responsible for the extinction of indigenous species, sustainable disturbance of ecosystems and economic damage. Consequently, there is increasing need and effort for the control and management of invasions. This requires the understanding of the mechanisms underlying the invasion process. Recently, many factors affecting the speed and the pattern of the spread of an introduced species have been identified, either empirically and/or theoretically, for example stochasticity (Lewis, 1997, 2000; Lewis and Pacala, 2000; Malchow et al., 2004), resource availability, spatial heterogeneity (Murray, 2003), environmental borders (Keitt et al., 2001), predation (Fagan and Bishop, 2000; Vinogradov et al., 2000; Owen and Lewis, 2001; Petrovskii et al., 2002a,b), competition (Okubo et al., 1989; Hart and Gardner, 1997), evolutionary changes (Lambrinos, 2004), large-scale phenomena such as weather conditions or (long-range) dispersal/transport effects (Hengeveld, 1989; Williamson, 1996; Shigesada and Kawasaki, 1997; Clark et al., 2001). For recent reviews, see Fagan et al. (2002), Case et al. (2005), Hastings et al. (2005), Holt et al. (2005) and Petrovskii et al. (2005).

Substantial populations of introduced alien species are reported to establish successfully, but then to collapse spontaneously (Simberloff and Gibbons, 2004). This phenomenon is also referred to as population crash. Pathogens are often suggested as an explanation, but in most cases there are neither documented empirical data nor identified pathogens.

Detailed theoretical studies of the influence of infectious diseases on the spread of species are still missing, but see Petrovskii et al. (2005), though there seems to be strong empirical evidence that pathogens and parasites play a major role in invasion biology, cf. Keane and Crawley (2002), Torchin et al. (2002), Clay (2003), Tompkins et al. (2003), Anderson et al. (2004), Lee and Klasing (2004), Prenter et al. (2004) and references therein.

The goal of this paper is to find conditions under which a subsequently released disease can be responsible for the decline, block or slow-down of established invasive species. Therefore, a two-component reaction-diffusion model describing the transmission of the pathogen and the spread of the host population is introduced in the next section. The host population exhibits a strong Allee effect (Dennis, 1989; Courchamp et al., 1999; Stephens et al., 1999; Stephens and Sutherland, 1999). The peculiar role of the Allee effect in biological invasions is increasingly acknowledged in recent studies (Lewis and Kareiva, 1993; Kot et al., 1996; Veit and Lewis, 1996; Grevstad, 1999; Takasu et al., 2000; Owen and Lewis, 2001; Wang and Kot, 2001; Garrett and Bowden, 2002; Petrovskii et al., 2002a,b; Shigesada and Kawasaki, 2002; Wang et al., 2002; Liebhold and Bascompte, 2003; Cappuccino, 2004; Davis et al., 2004; Drake, 2004; Leung et al., 2004; Petrovskii et al., 2005).

Reaction-diffusion models are essential and analytically tractable tools to understand invasion dynamics (Shigesada and Kawasaki, 1997; Okubo and Levin, 2001; Murray, 2003). This study is based upon travelling wave analysis (e.g. McKean, 1970; Aronson and Weinberger, 1975; Hadeler and Rothe, 1975; Fife and McLeod, 1977), in order to determine the front speeds which correspond to the asymptotic rate of spread. This approach can be traced back to the seminal works of Luther (1906), Fisher (1937), Kolmogorov et al. (1937) and Skellam

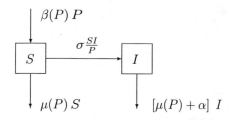

Figure 3.1: Transfer diagram of the infectious disease. More explanations can be found in the text.

(1951).

The outline of this paper is as follows. First, the model is described in section 3.2. Then, examples of different patterns of spread are given by way of numerical simulations, including both the slow-down of invasion and the population decline to ultimate extinction. The travelling wave analysis in section 3.4 reveals the conditions for the different behaviour. Finally, these results are discussed and related to similar work and a more general view of invasion patterns.

3.2   Mathematical model

We consider a spatiotemporal model for the spread of an invading species and its interplay with an introduced infectious disease. Therefore, the total population density $P = P(t, x) \geq 0$ at time t and spatial location x is split into a susceptible $S = S(t, x) \geq 0$ and an infected part $I = I(t, x) \geq 0$ with $P = S + I$. The epidemiological structure of the considered infectious disease is illustrated by the transfer diagram in Figure 3.1. Let $g(P)$ be the per-capita growth rate. We assume that the host population is subject to a strong Allee effect, say

$$g(P) = a(K_+ - P)(P - K_-) \,, \tag{3.1}$$

where $K_+ > 0$ is the carrying capacity and K_-, $0 < K_- < K_+$, is the minimum viable population density, below which the disease-free population goes extinct due to the Allee effect. The parameter $a > 0$ regulates the maximum growth rate. There is no recovery from the disease: Once a susceptible individual is exposed, it remains lifelong carrier of the pathogen. Transmission between susceptibles and infected is via the standard incidence, also called proportionate mixing or frequency-dependent transmission, cf. Nold (1980), Hethcote (2000) and McCallum et al. (2001). Infected individuals suffer an additional disease-related mortality $\alpha > 0$, also called the virulence. There is no vertical transmission, i.e., infected do not transmit the disease to their offspring. Thus, infected reproduce into the susceptible class, and the per-capita net growth rate is split into

$$g(P) = \beta(P) - \mu(P)$$

with $\beta(P) \geq 0$ being the fertility function and $\mu(P) \geq 0$ being the mortality function. We shall consider the following forms of $\beta(P)$ and $\mu(P)$, for which

(3.1) holds:

$$\beta(P) = a(-P^2 + [K_+ + K_- + e]P + c) , \qquad (3.2)$$
$$\mu(P) = a(eP + K_+K_- + c) , \qquad (3.3)$$

with parameters $c \geq 0$ and $e \geq 0$. The fertility function is quadratic and concave, i.e., $\beta(P)$ is increasing with P for low densities, reaching a maximum value at some intermediate density and decreasing for high densities. This may be due to biological reasons inducing the Allee effect, see the references in the introduction. The mortality function is assumed to increase linearly with the population density as it is usually assumed in logistic growth models. Note that with increasing values of the parameter e, the decreasing branch of the fertility function becomes less prominent in the range $P \in [0, K_+]$. In the mortality function, the effect of density dependence increases with e. The parameter c controls both vital functions in a density-independent way. Strictly speaking, equations (3.2-3.3) make sense as long as the fertility function remains nonnegative; this holds true at least for $0 \leq P \leq K_+ + K_- + e$ which is sufficient for this model. Together with diffusion as spatial propagation mechanism, we obtain a system of two partial differential equations (PDEs)

$$\frac{\partial S}{\partial t} = -\sigma \frac{SI}{P} + \beta(P)P - \mu(P)S + D_S \Delta S , \qquad (3.4)$$
$$\frac{\partial I}{\partial t} = \sigma \frac{SI}{P} - \mu(P)I - \alpha I + D_I \Delta I , \qquad (3.5)$$

where Δ is the Laplacian and $\sigma > 0$ the transmission coefficient. This model has been proposed and investigated by Hilker et al. (accepted) for equal diffusion coefficients $D = D_S = D_I$, because the disease is assumed not to affect the spreading behaviour. The same assumption will be used in this paper.

In their numerical simulations, they found that a spreading population can go extinct, when the disease is supplementarily introduced. This is a spatial phenomenon, because in the spatially homogeneous case both the host population and the disease coexist. This paper is concerned with an analytical explanation of this effect. Therefore, we restrict ourselves to one-dimensional space, i.e. $\boldsymbol{x} = x$, and introduce the dimensionless quantities

$$\tilde{S} = \frac{S}{K_+} , \qquad \tilde{t} = aK_+^2 t , \qquad \tilde{x} = K_+ \sqrt{\frac{a}{D}} x ,$$
$$\tilde{I} = \frac{I}{K_+} , \qquad \tilde{u} = \frac{K_-}{K_+} \in (0,1) , \qquad \tilde{\alpha} = \frac{\alpha}{aK_+^2} > 0 ,$$
$$\tilde{\sigma} = \frac{\sigma}{aK_+^2} > 0 , \qquad \tilde{c} = \frac{c}{K_+^2} \geq 0 , \qquad \tilde{e} = \frac{e}{K_+} \geq 0 .$$

Dropping the tildes for notational simplicity, we thus obtain for (3.4-3.5) the system

$$\frac{\partial S}{\partial t} = -\sigma \frac{SI}{P} + [(1 + u + e - P)P + c]P - (eP + u + c)S + \frac{\partial^2 S}{\partial x^2} , \quad (3.6)$$
$$\frac{\partial I}{\partial t} = \left[\sigma \frac{S}{P} - eP - u - c - \alpha \right] I + \frac{\partial^2 I}{\partial x^2} , \qquad (3.7)$$

where $P = S + I$ now denotes the dimensionless total population density. The number of parameters has been reduced from eight to five. System (3.6-3.7) is studied along with no-flux boundary conditions. Because we are interested in an invasion problem with a subsequently introduced disease in the wake of the invasion, we differentiate three regions of initial conditions. Without loss of generality, the right boundary is assumed to be still non-invaded. In the middle, there are only susceptibles. At the left boundary, the disease is assumed to have been introduced such that susceptibles and infected coexist:

$$S(0,x) \quad = \quad S_l > 0 \quad \text{if} \quad x < x_S \text{ and } 0 \text{ otherwise ,} \tag{3.8}$$

$$I(0,x) \quad = \quad I_l > 0 \quad \text{if} \quad x < x_I \text{ and } 0 \text{ otherwise ,} \tag{3.9}$$

with $x_S > x_I > 0$. There is a singularity in system (3.6-3.7) at the trivial solution. Introducing the prevalence $i = I/P \in [0,1]$, we reformulate (3.6-3.9) in (P,i) space:

$$\frac{\partial P}{\partial t} \quad = \quad [(1-P)(P-u) - \alpha i]P + \frac{\partial^2 P}{\partial x^2} , \tag{3.10}$$

$$\frac{\partial i}{\partial t} \quad = \quad [\sigma - \alpha - c - (1 + u + e - P)P - (\sigma - \alpha)i]i$$
$$+ \frac{2}{P}\frac{\partial P}{\partial x}\frac{\partial i}{\partial x} + \frac{\partial^2 i}{\partial x^2} , \tag{3.11}$$

$$P(0,x) = \begin{cases} S_l + I_l & \text{if } x < x_I , \\ S_l & \text{if } x_I \le x < x_S , \\ 0 & \text{if } x \ge x_S , \end{cases}$$

$$i(0,x) = \begin{cases} I_l/(S_l + I_l) & \text{if } x < x_I , \\ 0 & \text{if } x_I \le x < x_S , \\ i_r \in [0,1] & \text{if } x \ge x_S . \end{cases} \tag{3.12}$$

Note the convenient expressions in (3.10), but due to the transformation, (3.11) contains a term with the inverse of P and the spatial derivatives of both P and i. However, if we consider the spatially homogeneous version of (3.10-3.11), the singularity vanishes. The stationary states and their stabilities are summarized in Table 3.1. They have been found by Hilker et al. (accepted) for a generalized model with $\beta(P)$ being concave and $\mu(P)$ being nondecreasing and convex. There may be up to six equilibria. Periodic solutions are proven not to exist (Poincaré-Bendixson theory). The dynamics are bistable. If $\sigma - \alpha < c$, the disease cannot establish and the host population either goes extinct or approaches the carrying capacity, depending on the initial conditions. Further increasing the disease-related parameter combination $\sigma - \alpha$, the trivial solution becomes unstable and the disease-induced extinction state $(0, i_2 = 1 - \frac{c}{\sigma - \alpha})$ emerges. This equilibrium corresponds to extinction as a consequence of infection (de Castro and Bolker, 2005). While the host population tends to zero, the prevalence approaches $i_2 > 0$. Next, if $\sigma - \alpha > u + c + ue$, up to two coexistence states appear. The smaller one (P_{3-}, i_{3-}) is always unstable, while the larger one (P_{3+}, i_{3+}) changes stability with the disease-free carrying capacity state and is locally stable as soon as it exists. The nontrivial equilibria disappear if

(P^*,i^*)	c	$u+c+ue$	$u+c+e$	Eq.s (3.13-3.14) $\rightarrow \sigma-\alpha$	
$(0,0)$	l.a.s.	unstable	unstable	unstable	unstable
$(0,i_2)$	–	l.a.s.	l.a.s.	l.a.s.	g.a.s.
$(u,0)$	unstable	unstable	unstable	unstable	unstable
$(1,0)$	l.a.s.	l.a.s.	l.a.s.	unstable	unstable
(P_{3-},i_{3-})	–	–	unstable	unstable	–
(P_{3+},i_{3+})	–	–	–	l.a.s.	–

Table 3.1: Results of the stability analysis of the spatially homogeneous version of (3.10-3.11). The left column contains the stationary states (P^*,i^*). The other columns are divided according to the parameter regions along the ray for $\sigma - \alpha$ which are separated by the values in the top row. The most right column corresponds to the case that additionally equations (3.13-3.14) hold. "l.a.s." stands for locally asymptotically stable, "g.a.s." for globally asymptotically stable, and "–" means that the stationary state does not exist or is not feasible.

additionally

$$e > \frac{\sigma}{\alpha}(1-u)\,, \tag{3.13}$$

$$c < \frac{(\sigma-\alpha)(u+\alpha)}{\alpha} - \frac{[\sigma(1+u)+\alpha e]^2}{4\alpha\sigma}\,. \tag{3.14}$$

In this case the dynamics are monostable with a globally stable disease-induced extinction state. The total population densities of the coexistence states are

$$P_{3+,3-} = \frac{1}{2}\left[1+u+\frac{e\alpha}{\sigma}\right.$$
$$\left. \pm \sqrt{\left(1+u+\frac{e\alpha}{\sigma}\right)^2 - \frac{4}{\sigma}\left[(\sigma-\alpha)(u+\alpha)-c\alpha\right]}\,\right]. \tag{3.15}$$

They can be found as the roots of the two nontrivial, quadratic nullclines

$$i(P) = \frac{(1-P)(P-u)}{\alpha} = \frac{\sigma-\alpha-c-(1+u+e-P)P}{\sigma-\alpha}\,, \tag{3.16}$$

which are displayed, for example, in Figure 3.6a.

3.3 Slow-down and reversal of invasion fronts: numerical observations

When the invasion front of the host population is not influenced by the disease, for example if $i = 0$, system (3.10-3.11) reduces to a single-species model in which growth is described by a strong Allee effect. In this case, the invasion takes place via a travelling frontal wave with the unique speed

$$v = \frac{1}{\sqrt{2}}(1-2u)\,, \tag{3.17}$$

cf. Lewis and Kareiva (1993). If $u < 1/2$, the population front propagates in positive direction, while it runs backward otherwise.

In this paper, we assume that the population initially propagates forward, say $0 < u < 1/2$, and still continues spreading, when the disease is introduced in the wake of the invasion front of the host species, i.e.,

- we restrict to the parameter region in which the coexistence state is stable, cf. Table 3.1,

- the initial conditions prevent mere extinction due to the Allee effect, say (i) at least $P_l > u$ and (ii) the spatially critical threshold length is exceeded (Hilker et al., accepted), i.e., x_S is large enough.

This section is concerned with numerical simulations. The second summand on the right-hand side of (3.11) causes enormous numerical difficulties. Note that it can be interpreted as an advection term with velocity

$$-\frac{2}{P}\frac{\partial P}{\partial x} \, ,$$

i.e., the direction of the advection depends on the spatial gradient of P. The term does not play a role at the head and the wake of the fronts, because there at least one of the spatial derivatives vanishes. However, this term could become important, if the transition layers of both fronts for P and i overlap. Therefore, we consider for the numerical simulations system (3.6-3.9). In order to numerically handle the singularity, a threshold δ is applied, below which the transmission terms are set to zero. Throughout this paper, it will be kept as small as $\delta = 10^{-100}$. As will become evident below, the reason behind is that the extinction state $S = I = 0$ is stable. Experiences from numerical simulations elucidate that this extremely small choice of δ is necessary, in order to avoid artificial effects at the propagating wave front. Moreover, a sufficiently small time discretization is required.

Extensive numerical investigations indicate the emergence of travelling frontal waves. An example is given in Figure 3.2, where the numerical solutions are displayed in P and i state variables. At the head of the population front, the host species continues invasion in open space with speed v_1. The infection spreads into the susceptible population at carrying capacity with speed v_2 and settles towards the coexistence state (Figure 3.2a). If the prevalence front comes closer to the population front, there is a sudden boost of the prevalence in front of the population front (Figure 3.2b,c). Due to diffusion, there is a band of microscopic prevalence values moving ahead the actual prevalence front. Together with small total population densities, the system dynamics quickly tends to the disease-induced extinction state, because the trivial state is unstable. The travelling wave thus connects the coexistence state with the disease-induced extinction state, which has to establish from the trivial state. A too small choice of the threshold δ in the numerical scheme would affect an abrupt interruption of the rapid propagation of the disease-induced extinction state through the empty space (with speed v_3). Note that it is instructive to display the dynamics in P and i, because otherwise the decisive distinction between the unstable trivial state $(0, 0)$ and the stable disease-induced extinction state $(0, i_2)$ would be not as obvious.

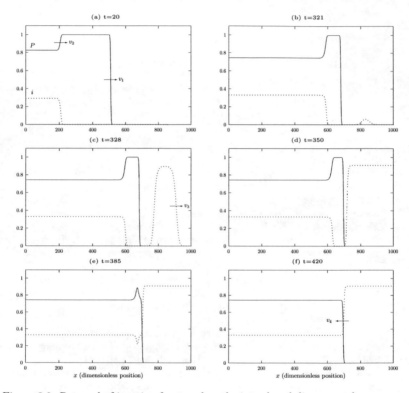

Figure 3.2: Reversal of invasion fronts, when the introduced disease catches up the front of the host population. The solid line represents P, the dotted line i. Solutions are obtained, as throughout this paper, by integrating (3.6-3.9) with the Runge-Kutta scheme of fourth order for local interactions, with the explicit Euler scheme for diffusion and applying a threshold as described in the text due to the singularity at the trivial state. Other parameter values: $\alpha = 0.5, \sigma = 1.6, c = 0.1, e = 0.5, u = 0.1, S_l = 1, I_l = 0.001, x_S = 500, x_I = 200$.

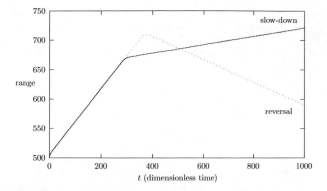

Figure 3.3: Range expansion of the total population for $\alpha = 0.3$ (slow-down) and $\alpha = 0.5$ (reversal). The range is the (dimensionless) distance which is occupied by the species with densities above some threshold, here set to 0.01. The slopes of the displayed lines correspond to the wave speeds. Other parameters as in Figure 3.2.

The appearance of the prevalence "hump" in front of the front is a transient effect of small densities, which has also been referred to as the "atto-fox problem" (Mollison, 1991). Since this is often regarded as artificial, note that the effect could be damped by simply applying an appropriate threshold, below which densities are set to zero, cf. Gurney et al. (1998), Gurney and Nisbet (1998) and Cruickshank et al. (1999). But the results would qualitatively remain the same, namely the disease-induced extinction state propagates to the right-hand side, while the coexistence state moves with speed v_4. In the shown example, the population front becomes recessive (Figure 3.2d-f), and the population ultimately goes extinct. For other parameter values, e.g. a smaller value of the virulence, the population front remains invasive, but with a slower wave speed, i.e. $v_4 < v_1$. This reversal and slow-down of the invasion fronts are illustrated in the range expansion diagram in Figure 3.3.

The next section is concerned with finding analytical conditions for either the slow-down or the reversal. This is achieved by deriving approximations for the wave speed v_4.

3.4 Travelling wave analysis

Although there are some results proving the existence of travelling wave solutions to systems similar to (3.6-3.9), e.g. Dunbar (1983, 1984, 1986), Hosono (1989), Huang et al. (2003) and Li et al. (submitted), we know of no proof of travelling wave solutions for this specific system. For the case that one species (predator) spreads into the range of another spreading species (prey), simulations by Sherratt et al. (1995, 1997) and Petrovskii and Malchow (1999, 2001) show that spatiotemporal dynamics can result in irregular oscillations – provided that the local dynamics fulfill some special conditions. Since these requirements differ from the properties of our model, we do not expect such a behaviour. As a result from the numerical simulations in the previous section, we consider travelling wave solutions to (3.10-3.12) of the form $\tilde{P}(z) = P(t, x)$ and $\tilde{i}(z) = i(t, x)$ with $z = x - vt$ and wave speed v. Omitting the tildes for notational simplicity and substituting yield the following system of ordinary differential equations of second order

$$0 = P'' + vP' + [(1 - P)(P - u) - \alpha i] P , \qquad (3.18)$$

$$0 = i'' + \left(v + \frac{2}{P}P'\right) i'$$
$$+ [\sigma - \alpha - (1 + u + e - P)P - c - (\sigma - \alpha)i] i , \qquad (3.19)$$

where the prime denotes differentiation with respect to z. Focusing on the situation lastly discussed in the previous section, cf. Figure 3.2f, we specify the boundary conditions as

$$P(-\infty) = P_{3+} , \qquad P(\infty) = 0 ,$$
$$i(-\infty) = i_{3+} , \qquad i(\infty) = i_2 .$$

Our main interest is in the direction of the travelling wave, say whether it moves to the left or to the right. Because of this we consider equation (3.18) for the total population density. By introducing $Q = P'$, this single equation of second order can be reduced to a system of two equations of first order, namely

$$P' = Q , \qquad (3.20)$$
$$Q' = -vQ - [(1 - P)(P - u) - \alpha i] P , \qquad (3.21)$$

with boundary conditions

$$P(-\infty) = P_{3+} , \qquad P(\infty) = 0 ,$$
$$Q(-\infty) = 0 , \qquad Q(\infty) = 0 .$$

Note that (3.21) depends on the prevalence i, which induces mathematical difficulties. In subsection 3.4.1, we will determine exact wave speeds for cases, in which we can make use of different time scale dynamics. This enables us to approximate the prevalence. In subsection 3.4.2, we show that we can also find good approximations for other cases.

3.4.1   Exact wave speed solutions for fast dynamics

Our model depends on five parameters, cf. (3.18-3.19). However, u is restricted to be within the interval $(0, 0.5)$, cf. section 3.3. Furthermore, the virulence α cannot be very large, because otherwise there are too many deaths due to the disease, and vital dynamics would become irrelevant. Thus, there remain the three parameters σ, e and c. In this subsection, we will investigate the cases in which the transmission coefficient σ becomes very large. This situation may arise when the disease is transmitted very fast. Moreover, the combinations of additionally large growth parameters e and/or c are also taken into account. In all these cases, a unique wave speed can be approximated analytically. The case that σ is small will be considered in the next subsection.

Firstly, we assume that both the disease transmission as well as the vital dynamics (births and deaths) occur on a fast time scale, i.e., we replace e, c and

σ by e/ε, c/ε and σ/ε, respectively, and then let $\varepsilon \to 0$. Consider equation (3.11) for i. Substituting e, c and σ as indicated and multiplying with ε yield

$$\varepsilon \frac{\partial i}{\partial t} = [\sigma - \varepsilon\alpha - c - (\varepsilon + \varepsilon u + e - \varepsilon P)P - (\sigma - \varepsilon\alpha)i]\, i + \varepsilon \frac{2}{P} \frac{\partial P}{\partial x} \frac{\partial i}{\partial x} + \varepsilon \frac{\partial^2 i}{\partial x^2} \ .$$

With $\varepsilon \to 0$, this reduces to

$$0 = [\sigma - c - eP - \sigma i]\, i \ ,$$

and thus $i = 0$, which corresponds to the disease-free states, or

$$i = 1 - \frac{c}{\sigma} - \frac{e}{\sigma} P \tag{3.22}$$

for cases in which the infection can establish. The total population density of the stable coexistence state reduces to

$$P_{3+} = \frac{1}{2}\left[1 + u + \frac{e\alpha}{\sigma} + \sqrt{\left(1 + u + \frac{e\alpha}{\sigma}\right)^2 - 4\left(u + \alpha - \frac{c\alpha}{\sigma}\right)} \right] \ . \tag{3.23}$$

Hence, the prevalence can be described by the linearly decreasing straight line (3.22), cf. Figure 3.4a. Incorporating this in (3.20-3.21) yields

$$P' = Q \ ,$$
$$Q' = -vQ - \left[(1-P)(P-u) - \alpha\left(1 - \frac{c}{\sigma} - \frac{e}{\sigma}P\right)\right]P \ .$$

We are looking for a heteroclinic connection between $(P,Q) = (P_{3+}, 0)$ and $(P,Q) = (0,0)$. Therefore, following Lewis and Kareiva (1993) and Murray (2002), we make the ansatz

$$P' = Q = AP(P_{3+} - P) \ .$$

Then

$$P'' = AP'(P_{3+} - 2P) = A^2 P(P_{3+} - P)(P_{3+} - 2P) \ ,$$

and substituting these expressions together with (3.22) in (3.18) yields

$$0 = \left[\underbrace{(2A^2 - 1)}_{c_2} P^2 + \underbrace{\left(1 + u + \frac{\alpha e}{\sigma} - vA - 3A^2 P_{3+}\right)}_{c_1} P \right.$$
$$\left. \underbrace{-u - \alpha + \frac{c\alpha}{\sigma} + vAP_{3+} + A^2 P_{3+}^2}_{c_0} \right] P \ .$$

This equation must hold for all $P \in [0, P_{3+}]$. For $P = 0$ it obviously does, and for all other values of P we require that $c_2 = c_1 = c_0 = 0$. From $c_2 = 0$ we obtain a condition for A, namely

$$A = \pm \frac{1}{\sqrt{2}} \ ,$$

and from the remaining two conditions $c_1 = c_0 = 0$ we obtain equations for the wave speed v. They must be equal, which they are, indeed, for the values of A as above and P_{3+} as in (3.23).

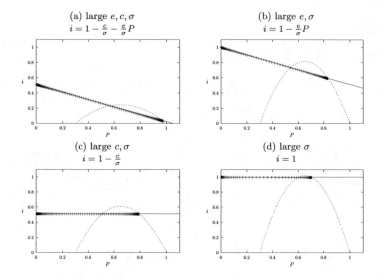

Figure 3.4: Solutions of the PDE system (3.6-3.9), restricted to the asymptotic propagation of the travelling fronts, displayed in (P, i) phase plane with cross points, for different cases of fast dynamics. The solid lines are the prevalence approximations as given in the respective panel. The dotted lines are the nullclines of the total population, cf. the left part of (3.16). Parameter values: (a) $e = c = 100.0, \sigma = 205.0, \alpha = 0.5$, (b) $e = 100.0, c = 0.1, \sigma = 205.0, \alpha = 0.15$, (c) $e = 100.0, c = 0.1, \sigma = 205.0, \alpha = 0.2$, (d) $e = c = 0.1, \sigma = 205.0, \alpha = 0.12$. All other values have been kept as $u = 0.3$, $S_l = S_{3+}, I_l = I_{3+}, x_S = 500, x_I = 200$.

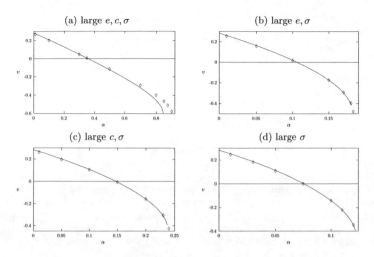

Figure 3.5: Analytically derived wave speeds (3.24-3.27) for the different combinations of fast transmission and vital dynamics, cf. Figure 3.4 (same parameter values). The points are numerically obtained results.

Because we are looking for a connection from $(P, P') = (P_{3+}, 0)$ to $(P, P') = (0, 0)$, P must decrease, which means that $P' \leq 0$ and that the trajectory is situated in the fourth quadrant. Hence, choosing $A = -1/\sqrt{2}$, we obtain the unique wave speed

$$v = \frac{\sqrt{2}}{4} \left[-1 - u - \frac{e\alpha}{\sigma} + 3\sqrt{\left(1 + u + \frac{e\alpha}{\sigma}\right)^2 - 4\left(u + \alpha - \frac{c\alpha}{\sigma}\right)} \right] . \qquad (3.24)$$

Figure 3.5a shows (3.24) in comparison with the numerically obtained values, demonstrating that the speeds match very well.

Next, we can repeat this procedure analogously for all other cases of fast dynamics, obtaining approximations for the prevalence as given and plotted in Figure 3.4b-d. Note that a large c corresponds to the situation, in which both the fertility as well as the mortality function occur on a fast time scale. For a large e, the effect of density dependence is increased. Hence, the dynamics become fast for high densities. The respective approximations of i are in all cases straight lines. If e is large, then they depend on P (Figure 3.4a-b). Otherwise, i can be approximated by a constant. For large values of c, this constant is below unity (Figure 3.4c), otherwise it is exactly unity (Figure 3.4d). The corresponding wave speeds are

$$v = \frac{\sqrt{2}}{4}\left[-1 - u - \frac{e\alpha}{\sigma} + 3\sqrt{\left(1 + u + \frac{e\alpha}{\sigma}\right)^2 - 4(u + \alpha)}\right] \quad \text{(large } e,\sigma) \,, \quad (3.25)$$

$$v = \frac{\sqrt{2}}{4}\left[-1 - u + 3\sqrt{(1 + u)^2 - 4\left(u + \alpha - \frac{c\alpha}{\sigma}\right)}\right] \quad \text{(large } c,\sigma) \,, \quad (3.26)$$

$$v = \frac{\sqrt{2}}{4}\left[-1 - u + 3\sqrt{(1 - u)^2 - 4\alpha}\right] \quad \text{(large } \sigma) \,. \quad (3.27)$$

The plots in Figures 3.5b-d underline the good approximation of the wave speeds obtained in numerical simulations. The analytic solutions reveal that the waves exist, i.e., v is a real number, as soon as the stable coexistence states exist in the corresponding situations, cf. for instance (3.23). Note that v can become negative. In this case, there is a reversal of the invasion front. A zero-velocity corresponds to the stop of the invasion. Though such a stationary front is superimposed by noisy fluctuations in nature and thus unlikely to be observed, this can be an explanation for at least transient species borders in spatially homogeneous environments, cf. Holt et al. (2005) and references therein. In the case that exclusively σ is large, the speed only depends on u and α. In the other cases, the speed additionally depends on e, c and σ. To be more precise, just the ratios of e/σ and c/σ influence the rate of spread. Note that in all the cases of fast dynamics, the prevalence is approximated by its nullcline of the non-spatial model. This is due to the fact that the disease transmission occurs on the fast time scale.

3.4.2   Approximations for small σ

If σ is small, which corresponds to a slow transmission of the disease, while the vital dynamics occur on a fast time scale, i.e., one or both of the parameters e and c are large, then the disease cannot establish. This means, that there is no coexistence state, cf. the parameter regions in Table 3.1. In the last remaining situation neither the disease transmission nor the vital dynamics occur on a fast time scale. In this case, the prevalence cannot be approximated as before. However, experiences from numerical simulations indicate that, nonetheless, an estimate of the unique wave speed can be obtained.

Consider for example the dynamics in Figure 3.6a, which corresponds to the situation shown in Figure 3.2. There is still an apparent similarity between the PDE solution and the nullcline of the prevalence, cf. (3.16), which makes us to try to approximate i in this way. I.e., we incorporate the ansatz

$$i = 1 - \frac{c}{\sigma - \alpha} - \frac{1 + u + e}{\sigma - \alpha}P + \frac{1}{\sigma - \alpha}P^2 \,,$$

cf. (3.16), in equation (3.21) and proceed as in subsection 3.4.1, thus obtaining the unique wave speed

$$v = \sqrt{\frac{2(\sigma - \alpha)}{\sigma}}\left[\frac{3\sigma}{2(\sigma - \alpha)}P_{3+} - 1 - u - \alpha\frac{1 + u + e}{\sigma - \alpha}\right] \,, \quad (3.28)$$

with P_{3+} as in (3.15). This analytical approximation is plotted against the virulence in Figure 3.6b (dotted line). The predicted wave speeds closely match

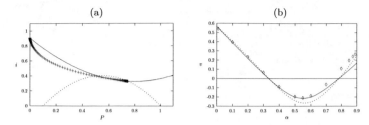

Figure 3.6: (a) Asymptotic front solution of the PDE system (3.6-3.9), displayed in (P, i) phase space with cross points. The solid and the dotted line are the nullclines of i and P, respectively. (b) Analytically approximated wave speeds (3.28) in dotted line and (3.30) in solid line. The points are numerically obtained results. Parameter values as in Figure 3.2.

the numerically observed ones. Only for intermediate values of α, there are slight underestimations. Note that the wave speed becomes positive again for large virulences, i.e., the population front propagates forward again.

There are, however, situations in which the PDE solution cannot be approximated by the nullcline, and these can naturally be assumed to be the more general case. An example is given in Figure 3.7a: The PDE solution would suggest rather a singular perturbation analysis than an approximation by the nullcline. Yet another approach shall be guided now. We seek for another approximation of the term $(1 - P)(P - u) - \alpha i$ in equation (3.18). Let us try

$$(1 - P)(P - u) - \alpha i = (P_{3+} - P)(P - P_{3-}) , \qquad (3.29)$$

of which we know that this clearly holds for $i = P_{3+}$ and $i = P_{3-}$. Note that this ansatz corresponds to the case if i would be a constant (as for example in the single-species models in Murray (2002, 2003)). Figure 3.7b shows both the left-hand side and the right-hand side of (3.29) with values of i from the numerical front solution. They are in good accordance for all $P \in [0, P_{3+}]$. Hence, let us make use of this approximation and substitute the term $(1 - P)(P - u) - \alpha i$ in (3.21) by the right-hand side of (3.29). Then we obtain

$$\begin{aligned} P' &= Q , \\ Q' &= -vQ - (P_{3+} - P)(P - P_{3-})P . \end{aligned}$$

Finally, we yield the following simple expression for the unique wave speed

$$v = \frac{1}{\sqrt{2}} [P_{3+} - 2P_{3-}] . \qquad (3.30)$$

As we can easily see, the wave exists if the coexistence states exist. An invasion front is predicted to be reversed, if the stable coexistence state is not at least twice larger than the unstable one. The spread rate is always smaller than (3.17), because $0 < u < P_{3-} < P_{3+} < 1$. This means that the introduction of the disease slows down the host population. In Figure 3.7c, (3.30) is plotted against α and compared with the numerically obtained wave speeds, demonstrating a good accordance. This parameter setting is an example in which the wave

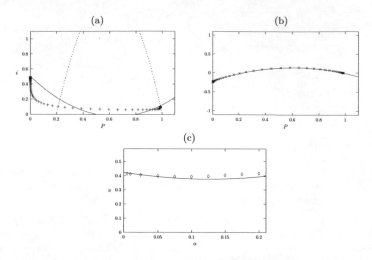

Figure 3.7: (a) Asymptotic front solution of the PDE system (3.6-3.9), displayed in (P, i) phase plane with cross points. The solid and the dotted line are the nullcline of i and P, respectively. (b) Approximation of the term $(1 - P)(P - u) - \alpha i$ (crosses) by the right-hand side of (3.29) (solid line). The data for i stem from the front solution of the PDE system. (c) Approximated wave speed (3.30) in solid line and numerically calculated data (points). Parameter values: $u = 0.2, e = 0.1, c = 0.4, \alpha = 0.1, \sigma = 0.9, S_l = S_{3+}, I_l = I_{3+}, x_S = 100, x_I = 50$.

speeds do not become negative for feasible α. Actually, equation (3.30) can also be applied to the parameter setting in Figure 3.6. The resulting wave speed approximation is plotted in Figure 3.6b as well (in solid line). Compared with (3.28), (3.30) fits very accurately for intermediate values of α, while there is a slight underestimatation for large values. However, both estimates yield the same threshold values of the virulence for the transitions from an invasvive to a recessive wave and vice versa.

The wave speeds for the fast dynamics as summarized in subsection 3.4.1 can be obtained as special cases of both (3.28) and (3.30) by simply letting the respective parameters become large. All these wave speed approximations as well as the disease-free wave speed (3.17), are particular cases of the fundamental setting with cubic nonlinearities in the interaction term of the PDE. Say, if r_0, r_1, and r_2 denote the roots of this polynomial, then the wave speed is given by

$$v = \frac{1}{\sqrt{2}} [r_2 - 2r_1 + r_0] \ .$$

In the case of (3.30) we have $r_2 = P_{3+}, r_1 = P_{3-}$ and $r_0 = 0$.

Finally, we are interested in the impact of the Allee threshold u on the reversal dynamics. In Figure 3.8, the predicted wave speeds (3.30) are plotted against both the virulence and the Allee threshold. For fixed values of u, there is a minimum wave speed for some intermediate virulence α. When the Allee threshold is varied, the wave speed decreases with u. This can also easily be checked by looking at the wave speed equations and their dependence on u. The

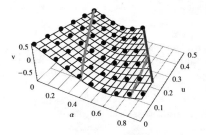

Figure 3.8: Dependence of the analytic wave speed solution (3.30) on the Allee threshold u and the virulence α. The points are numerically calculated wave speeds. The grey lines are isolines of zero-wavespeed and indicate transition from a recessive to an invasive wave. The thick black line delineates the feasible parameter region in which the coexistence state exists. Other parameters chosen as in Figure 3.2.

parameter regions, in which a front reversal is possible, gets larger with increasing Allee threshold. Recall that for $u > 0.5$ the disease-free host population wave would be recessive anyway.

3.5   Discussion and conclusions

In this paper, we have considered a reaction-diffusion model describing the spatiotemporal spread of a species which is governed by a strong Allee effect as well as an infectious disease. The invasion takes place via a travelling frontal wave. It has been shown that this invasion front of the host population can be slowed down, stopped or reversed by subsequently introduced pathogens. The invasion reversal and thus the ultimate population extinction depend on the virulence. When the additional disease-related mortality is within a certain range, the deaths due to infection overbalance the growth at the head of the population front.

The asymptotic rate of propagation decreases with an increasing strength of the Allee effect, i.e. a larger threshold u. This agrees with the results from single-species models, in which Allee effects slow down travelling wave solutions of reaction-diffusion (Lewis and Kareiva, 1993; Wang and Kot, 2001) and integrodifference equations (Wang et al., 2002). The possible reversal of invasion fronts has also been found by Owen and Lewis (2001) in a predator-prey model where the prey exhibits an Allee effect. Predator-prey and epidemiological SI models obey some structural analogy. Critical parameters for the reversal of the fronts are the predator mortality and the disease-related mortality, respectively. In both the predator-prey and the infectious disease model, the local models with neglected diffusion would predict coexistence. Recessive waves ("waves of extinction") have also been observed by Lewis and van den Driessche (1993) in a model of sterile insect release, in order to control an insect invasion. It would be interesting to know, what happens in a competition or any other model with negative impact on the focal species, when the latter is subject to an Allee effect.

This study has shown that reversal is possible for Allee thresholds u even approaching zero. Owen and Lewis (2001) have oberserved in their examples critical thresholds of about $u = 0.4$ for prey-predator interactions of simple Lotka-Volterra type and of about $u = 0.3$ for prey-predator interactions of Holling type III. Future work will have to consider different transmission functions. For instance, considering mass action transmission could also investigate whether the reversal depends on the disease-induced extinction state. Since the recessive waves are possible in prey-predator systems with Lotka-Volterra interaction, we would expect them to occur also in epidemiological models with transmission dynamics other than the standard incidence.

Owen and Lewis (2001) assumed that the predator diffuses much faster than the prey species, which, from a mathematical point of view, allows the application of singular perturbation theory. Here, we have considered equal diffusion coefficients of susceptibles and infected. The assumptions that some dynamical components are fast allowed to tract the problem analytically. Alternative approaches were also proposed which are based upon numerically observed features to guide an ansatz for deriving the wave speed expression as well. The diffusion coefficient of the infected could also be assumed to be reduced as a result of the infection. Hence, the diffusion coefficients would be unequal but still of the same order of magnitude. The analysis based on the fast dynamics assumptions should also hold in this case. Whether the alternative approaches turn out to be similarly accurate, would have to be checked in respective extensive numerical simulations.

A prerequisite of the reversal, however, is that one of the species is still invading into open space, while the other one is able to catch up this front. This can be regarded to be often the case, because invasion usually takes place by introduction of a small number of individuals. When these are disease-free, the pathogen is left behind and not present in the new habitat before another introduction of infected individuals (Pimentel, 1986).

Population crashes of alien invaders cannot be caused by indigenous pathogens, because they would have prevented a successful establishment. Mutational effects are more unlikely than the subsequent release of native pathogens, since evolution tends to coexistence states rather than to extinction, e.g. Roughgarden (1975), Ewald (1994), Yamamura (1996) and Boots and Sasaki (2003). Moreover, recent quantitative studies of Mitchell and Power (2003) and Torchin et al. (2003) demonstrate that invasive species, both plants and animals, appear to suffer from fewer pathogens and parasites in their new habitat than in their native range.

There are a lot of models dealing with the control of epidemic spread or investigating the impact of infectious diseases on non-native species, e.g. the control of rabies, see Murray (2003) and references therein, and immunocontraception (McCallum, 1996; Courchamp and Cornell, 2000; Suppo et al., 2000). Usually, the density of the host species is only reduced due to the disease, but, however, the population does not go extinct spatially. Hence, this phenomenon can be ascribed to the Allee effect, which is another evidence that Allee dynamics are a source of complex spatiotemporal dynamics (Lewis and Kareiva, 1993; Owen and Lewis, 2001; Petrovskii et al., 2002a,b, 2005).

The results of this study clearly highlight the importance and role of infectious diseases in invasion processes. Furthermore, they indicate that the introduction of natural pathogens may be applicable as a biological control method, in order to eradicate a pest species or at least to slow down its spread. This depends on the availability of appropriately harmful diseases and the range of their release. However, possible adverse effects to the ecosystem such as transmission to indigenous species have to be taken into account, of course.

Acknowledgement

F.M.H. and H.M. gratefully acknowledge support by the Japan Society for the Promotion of Science (Predoc and Research Fellowships PE04533 and S04716, respectively), which has made possible their stays in Japan during which parts of the presented results have been obtained.

F.M.H. is thankful for discussion with the groups of H. Seno, N. Shigesada, Y. Iwasa, N. Yamamura and Y. Hosono as well as with S.V. Petrovskii.

M.A.L. gratefully acknowledges support from a Canada Research Chair, NSERC Collaborative Research Opportunity and Discovery grants and the National Science Foundation grant no. DEB-0213698.

The autors thank two anonymous reviewers for helpful comments.

References

Anderson, P. K., Cunningham, A. A., Patel, N. G., Morales, F. J., Epstein, P. R., Daszak, P. (2004). Emerging infectious diseases of plants: pathogen pollution, climate change and agrotechnology drivers. *Trends in Ecology & Evolution* 19(10), 535–544.

Aronson, D. G., Weinberger, H. F. (1975). Nonlinear diffusion in population genetics, combustion, and nerve propagation. In *Partial differential equations and related topics* (Goldstein, J. A., ed.). No. 446 in Lecture Notes in Mathematics, Springer-Verlag, Berlin, pp. 5–49.

Boots, M., Sasaki, A. (2003). Parasite evolution and extinctions. *Ecology Letters* 6, 176–182.

Cappuccino, N. (2004). Allee effect in an invasive alien plant, pale swallow-wort *Vincetoxicum rossicum* (Asclepiadaceae). *Oikos* 106, 3–8.

Case, T. J., Holt, R. D., McPeek, M. A., Keitt, T. H. (2005). The community context of species' borders: ecological and evolutionary perspectives. *Oikos* 108, 28–46.

Clark, J. S., Lewis, M., Horvath, L. (2001). Invasion by extremes: population spread with variation in dispersal and reproduction. *The American Naturalist* 157(5), 537–554.

Clay, K. (2003). Parasites lost. *Nature* 421, 585–586.

Courchamp, F., Clutton-Brock, T., Grenfell, B. (1999). Inverse density dependence and the Allee effect. *Trends in Ecology & Evolution* 14(10), 405–410.

Courchamp, F., Cornell, S. J. (2000). Virus-vectored immunocontraception to control feral cats on islands: a mathematical model. *Journal of Applied Ecology* 37, 903–913.

Cruickshank, I., Gurney, W. S. C., Veitch, A. R. (1999). The characteristics of epidemics and invasions with thresholds. *Theoretical Population Biology* 56, 279–292.

Davis, H. G., Taylor, C. M., Civille, J. C., Strong, D. R. (2004). An Allee effect at the front of a plant invasion: *Spartina* in a Pacific estuary. *Journal of Animal Ecology* 92, 321–327.

de Castro, F., Bolker, B. (2005). Mechanisms of disease-induced extinction. *Ecology Letters* 8, 117–126.

Dennis, B. (1989). Allee effects: population growth, critical density, and the chance of extinction. *Natural Resource Modeling* 3, 481–538.

Drake, J. M. (2004). Allee effects and the risk of biological invasion. *Risk analysis* 24(4), 795–802.

Dunbar, S. R. (1983). Travelling wave solutions of diffusive Lotka-Volterra equations. *Journal of Mathematical Biology* 17, 11–32.

Dunbar, S. R. (1984). Travelling wave solutions of diffusive Lotka-Volterra equations: a heteroclinic connection in R^4. *Transactions of the American Mathematical Society* 286, 557–594.

Dunbar, S. R. (1986). Travelling waves in diffusive predator-prey equations: periodic orbits and point-to-periodic heteroclinic orbits. *SIAM Journal on Applied Mathematics* 46, 1057–1078.

Ewald, P. W. (1994). *Ecology of infectious diseases*. Oxford University Press, Oxford.

Fagan, W. F., Bishop, J. G. (2000). Trophic interactions during primary succession: herbivores slow a plant reinvasion at Mount St. Helens. *The American Naturalist* 155, 238–251.

Fagan, W. F., Lewis, M. A., Neubert, M. G., van den Driessche, P. (2002). Invasion theory and biological control. *Ecology Letters* 5, 148–158.

Fife, P. C., McLeod, J. B. (1977). The approach of solutions of nonlinear diffusion equations to travelling wave solution. *Archive for Rational Mechanics and Analysis* 65, 335–361.

Fisher, R. A. (1937). The wave of advance of advantageous genes. *Annals of Eugenics* 7, 355–369.

Garrett, K. A., Bowden, R. L. (2002). An Allee effect reduces the invasive potential of *Tilletia indica*. *Phytopathology* 92(11), 1152–1159.

Grevstad, F. S. (1999). Factors influencing the change of population establishment: implications of release strategies in biocontrol. *Ecological Applications* 9(4), 1439–1447.

Gurney, W. S. C., Nisbet, R. M. (1998). *Ecological dynamics*. Oxford University Press, New York.

Gurney, W. S. C., Veitch, A. R., Cruickshank, I., McGeachin, G. (1998). Circles and spirals: population persistence in a spatially explicit predator-prey model. *Ecology* 79(7), 2516–2530.

Hadeler, K. P., Rothe, F. (1975). Travelling fronts in nonlinear diffusion equations. *Journal of Mathematical Biology* 2, 251–263.

Hart, D. R., Gardner, R. H. (1997). A spatial model for the spread of invading organisms subject to competition. *Journal of Mathematical Biology* 35(8), 935–948.

Hastings, A., Cuddington, K., Davies, K. F., Dugaw, C. J., Elmendorf, S., Freestone, A., Harrison, S., Holland, M., Lambrinos, J., Malvadkar, U., Melbourne, B. A., Moore, K., Taylor, C., Thomson, D. (2005). The spatial spread of invasions: new developments in theory and evidence. *Ecology Letters* 8, 91–101.

Hengeveld, R. (1989). *Dynamics of biological invasions*. Chapman and Hall, London.

Hethcote, H. W. (2000). The mathematics of infectious diseases. *SIAM Review* 42(4), 599–653.

Hilker, F. M., Langlais, M., Petrovskii, S. V., Malchow, H. (accepted). A diffusive SI model with Allee effect and application to FIV. *Mathematical Biosciences* .

Holt, R. D., Keitt, T. H., Lewis, M. A., Maurer, B. A., Taper, M. L. (2005). Theoretical models of species' borders: single species approaches. *Oikos* 108, 18–27.

Hosono, Y. (1989). Singular perturbation analysis of travelling waves for diffusive Lotka-Volterra competitive models. In *Numerical and applied mathematics* (Brezinski, C., ed.). Baltzer, Basel, pp. 687–692.

Huang, J., Lu, G., Ruan, S. (2003). Existence of traveling wave solutions in a diffusive predator-prey model. *Journal of Mathematical Biology* 46(2), 132–152.

Keane, R. M., Crawley, M. J. (2002). Exotic plant invasions and the enemy release hypothesis. *Trends in Ecology & Evolution* 17(4), 164–170.

Keitt, T. H., Lewis, M. A., Holt, R. D. (2001). Allee effects, invasion pinning, and species' borders. *The American Naturalist* 157, 203–216.

Kolmogorov, A. N., Petrovskii, I. G., Piskunov, N. S. (1937). Étude de l'equation de la diffusion avec croissance de la quantité de matière et son application à un problème biologique. *Bulletin Université d'Etat à Moscou, Série internationale, section A* 1, 1–25.

Kot, M., Lewis, M. A., van den Driessche, P. (1996). Dispersal data and the spread of invading organisms. *Ecology* 77(7), 2027–2042.

Lambrinos, J. G. (2004). How interactions between ecology and evolution influence contemporary invasion dynamics. *Ecology* 85(8), 2061–2070.

Lee, K. A., Klasing, K. C. (2004). A role for immunology in invasion biology. *Trends in Ecology & Evolution* 19(10), 523–529.

Leung, B., Drake, J. M., Lodge, D. M. (2004). Predicting invasions: propagule pressure and the gravity of Allee effects. *Ecology* 85(6), 1651–1660.

Lewis, M. A. (1997). Variability, patchiness, and jump dispersal in the spread of an invading population. In *Spatial ecology. The role of space in population dynamics and interspecific interactions* (Tilman, D., Kareiva, P., eds.). Princeton University Press, Princeton, pp. 46–69.

Lewis, M. A. (2000). Spread rate for a nonlinear stochastic invasion. *Journal of Mathematical Biology* 41, 430–454.

Lewis, M. A., Kareiva, P. (1993). Allee dynamics and the spread of invading organisms. *Theoretical Population Biology* 43, 141–158.

Lewis, M. A., Pacala, S. (2000). Modeling and analysis of stochastic invasion processes. *Journal of Mathematical Biology* 41, 387–429.

Lewis, M. A., van den Driessche, P. (1993). Waves of extinction from sterile insect release. *Mathematical Biosciences* 116, 221–247.

Li, B., Weinberger, H. F., Lewis, M. A. (submitted). Existence of traveling waves for discrete and continuous time cooperative systems.

Liebhold, A., Bascompte, J. (2003). The Allee effect, stochastic dynamics and the eradication of alien species. *Ecology Letters* 6, 133–140.

Luther, R. (1906). Räumliche Ausbreitung chemischer Reaktionen. *Zeitschrift für Elektrochemie* 12, 596–600.

Malchow, H., Hilker, F. M., Petrovskii, S. V., Brauer, K. (2004). Oscillations and waves in a virally infected plankton system: Part I: The lysogenic stage. *Ecological Complexity* 1(3), 211–223.

McCallum, H. (1996). Immunocontraception for wildlife population control. *Trends in Ecology & Evolution* 11(12), 491–493.

McCallum, H., Barlow, N., Hone, J. (2001). How should pathogen transmission be modelled? *Trends in Ecology & Evolution* 16(6), 295–300.

McKean, H. P. (1970). Nagumo's equation. *Advances in Mathematics* 4, 209–223.

Mitchell, C. E., Power, A. G. (2003). Release of invasive plants from fungal and viral pathogens. *Nature* 421, 625–627.

Mollison, D. (1991). Dependence of epidemics and populations velocities on basic parameters. *Mathematical Biosciences* 107, 255–287.

Murray, J. D. (2002). *Mathematical biology. I: An introduction.* Springer-Verlag, Berlin, 3rd edition.

Murray, J. D. (2003). *Mathematical biology. II: Spatial models and biomedical applications.* Springer-Verlag, Berlin, 3rd edition.

Nold, A. (1980). Heterogeneity in disease-transmission modeling. *Mathematical Biosciences* 52, 227–2402.

Okubo, A., Levin, S. A. (2001). *Diffusion and ecological problems: Modern perspectives*. Springer-Verlag, New York, 2nd edition.

Okubo, A., Maini, P. K., Williamson, M. H., Murray, J. D. (1989). On the spatial spread of the gray squirrel in Britain. *Proceedings of the Royal Society of London B* 238, 113–125.

Owen, M. R., Lewis, M. A. (2001). How predation can slow, stop or reverse a prey invasion. *Bulletin of Mathematical Biology* 63, 655–684.

Petrovskii, S. V., Malchow, H. (1999). A minimal model of pattern formation in a prey-predator system. *Mathematical and Computer Modelling* 29, 49–63.

Petrovskii, S. V., Malchow, H. (2001). Wave of chaos: new mechanism of pattern formation in spatio-temporal population dynamics. *Theoretical Population Biology* 59(2), 157–174.

Petrovskii, S. V., Malchow, H., Hilker, F. M., Venturino, E. (2005). Patterns of patchy spread in deterministic and stochastic models of biological invasion and biological control. *Biological Invasions* 7, 771–793.

Petrovskii, S. V., Morozov, A. Y., Venturino, E. (2002a). Allee effect makes possible patchy invasion in a predator-prey system. *Ecology Letters* 5, 345–352.

Petrovskii, S. V., Vinogradov, M. E., Morozov, A. Y. (2002b). Formation of the patchiness in the plankton horizontal distribution due to biological invasion in a two-species model with account for the Allee effect. *Oceanology* 42(3), 363–372.

Pimentel, D. (1986). Biological invasions of plants and animals in agriculture and forestry. In *Ecology of biological invasions of North America and Hawaii* (Mooney, H. A., Drake, J. A., eds.). Springer-Verlag, New York, pp. 149–162.

Prenter, J., MacNeil, C., Dick, J. T. A., Dunn, A. M. (2004). Roles of parasites in animal invasions. *Trends in Ecology & Evolution* 19(7), 385–390.

Roughgarden, J. (1975). Evolution of marine symbiosis – a simple cost-benefit model. *Ecology* 56, 1201–1208.

Sherratt, J. A., Eagan, B. T., Lewis, M. A. (1997). Oscillations and chaos behind predator-prey invasion: mathematical artifact or ecological reality? *Philosophical Transactions of the Royal Society of London B* 352, 21–38.

Sherratt, J. A., Lewis, M. A., Fowler, A. C. (1995). Ecological chaos in the wake of invasion. *Proceedings of the National Academy of Sciences of the United States of America* 92, 2524–2528.

Shigesada, N., Kawasaki, K. (1997). *Biological invasions: Theory and practice*. Oxford University Press, Oxford.

Shigesada, N., Kawasaki, K. (2002). Invasion and the range expansion of species: effects of long-distance dispersal. In *Dispersal ecology* (Bullock, J., Kenward, R., Hails, R., eds.). Blackwell Science, Malden MA, pp. 350–373.

Simberloff, D., Gibbons, L. (2004). Now you see them, now you don't! - Population crashes of established introduced species. *Biological Invasions* 6, 161–172.

Skellam, J. G. (1951). Random dispersal in theoretical populations. *Biometrika* 38, 196–218.

Stephens, P. A., Sutherland, W. J. (1999). Consequences of the Allee effect for behaviour, ecology and conservation. *Trends in Ecology & Evolution* 14(10), 401–405.

Stephens, P. A., Sutherland, W. J., Freckleton, R. P. (1999). What is the Allee effect? *Oikos* 87(1), 185–190.

Suppo, C., Naulin, J.-M., Langlais, M., Artois, M. (2000). A modelling approach to vaccination and contraception programmes for rabies control in fox populations. *Proceedings of the Royal Society of London B* 267, 1575–1582.

Takasu, F., Yamamoto, N., Kawasaki, K., Togashi, K., Kishi, Y., Shigesada, N. (2000). Modeling the expansion of an introduced tree disease. *Biological Invasions* 2, 141–150.

Tompkins, D. M., White, A. R., Boots, M. (2003). Ecological replacement of native red squirrels by invasive greys driven by disease. *Ecology Letters* 6, 189–196.

Torchin, M. E., Lafferty, K. D., Dobson, A. P., McKenzle, V. J., Kuris, A. M. (2003). Introduced species and their missing parasites. *Nature* 421, 628–630.

Torchin, M. E., Lafferty, K. D., Kuris, A. M. (2002). Parasites and marine invasions. *Parasitology* 124, S137–S151.

Veit, R. R., Lewis, M. A. (1996). Dispersal, population growth, and the Allee effect: Dynamics of the House Finch invasion of eastern North America. *The American Naturalist* 148(2), 255–274.

Vinogradov, M. E., Shushkina, E. A., Anochina, L. L., Vostokov, S. V., Kucheruk, N. V., Lukashova, T. A. (2000). Mass development of ctenophore Beroe ovata Eschscholtz off the north-eastern coast of the Black Seas. *Oceanology* 40, 52–55.

Wang, M.-H., Kot, M. (2001). Speeds of invasion in a model with strong or weak Allee effects. *Mathematical Biosciences* 171(1), 83–97.

Wang, M.-H., Kot, M., Neubert, M. G. (2002). Integrodifference equations, Allee effects, and invasions. *Journal of Mathematical Biology* 44, 150–168.

Williamson, M. (1996). *Biological invasions*. Chapman & Hall, London.

Yamamura, N. (1996). Evolution of mutualistic symbiosis: a differential equation model. *Researches on Population Ecology* 38(2), 211–218.

Part II

Infections in
prey-predator systems

Chapter 4

Oscillations and waves in a virally infected plankton system Part I: The lysogenic stage

Horst Malchow[1], Frank M. Hilker[1], Sergei V. Petrovskii[2] and Klaus Brauer[1]

[1] Institute of Environmental Systems Research, Department of Mathematics and Computer Science, University of Osnabrück, 49069 Osnabrück, Germany

[2] Shirshov Institute of Oceanology, Russian Academy of Sciences, Nakhimovsky Prospekt 36, Moscow 117218, Russia

Published in *Ecological Complexity* 1(3), 211-223 (2004).[1]

Abstract

A model of phytoplankton-zooplankton dynamics is considered for the case of lysogenic viral infection of the phytoplankton population. The phytoplankton population is split into a susceptible (S) and an infected (I) part. Both parts grow logistically, limited by a common carrying capacity. Zooplankton (Z) is grazing on susceptibles and infected. The local analysis of the S-I-Z differential equations yields a number of stationary and/or oscillatory regimes and their combinations. Correspondingly interesting is the spatiotemporal behaviour, modelled by deterministic and stochastic reaction-diffusion equations. Spatial spread or suppression of infection will be presented just as well as competition of concentric and/or spiral population waves with non-oscillatory subpopulations for space. The external noise can enhance the survival and spread of susceptibles and infected, respectively, that would go extinct in a deterministic environment.

---

[1]Reprinted with kind permission from Elsevier. Copyright Elsevier B.V. 2004.

4.1   Introduction

Numerous papers have been published about pattern formation and chaos in
minimal prey-predator models of phytoplankton-zooplankton dynamics (Scheffer, 1991a; Malchow, 1993; Pascual, 1993; Truscott and Brindley, 1994; Malchow,
1996, 2000b; Malchow et al., 2001, 2004b). Different routes to local and spatiotemporal chaos (Scheffer, 1991b; Kuznetsov et al., 1992; Rinaldi et al., 1993;
Sherratt et al., 1995; Scheffer et al., 1997; Steffen et al., 1997; Petrovskii and Malchow, 1999, 2001; Malchow et al., 2002), diffusion- and differential-flow-induced
standing and travelling waves (Malchow, 1993; Menzinger and Rovinsky, 1995;
Malchow, 2000a; Satnoianu and Menzinger, 2000; Satnoianu et al., 2000; Malchow et al., 2003) as well as target patterns and spiral waves (Medvinsky et al.,
2000, 2002) have been found. Also the impact of external noise on patchiness
and transitions between alternative stable population states has been studied
(Steele and Henderson, 1992; Malchow et al., 2002; Sarkar and Chattopadhayay,
2003; Malchow et al., 2004a).

Much less than on plankton patchiness and blooming is known about marine viruses and their role in aquatic ecosystems and the species that they infect,
for reviews cf. Fuhrman (1999); Suttle (2000) as well as Wommack and Colwell (2000). There is some evidence that viral infection might accelerate the
termination of phytoplankton blooms (Jacquet et al., 2002).

Also the understanding of the importance of lysogeny is just at the beginning (Wilcox and Fuhrman, 1994; Jiang and Paul, 1998; McDaniel et al., 2002;
Ortmann et al., 2002). Contrary to lytic infections with destruction and without reproduction of the host cell, lysogenic infections are a strategy whereby
viruses integrate their genome into the host's genome. As the host reproduces
and duplicates its genome, the viral genome reproduces, too.

Mathematical models of the dynamics of virally infected phytoplankton populations are rare as well. The already classical publication is by Beltrami and
Carroll (1994), more recent work is of Chattopadhyay and Pal (2002) and Chattopadhyay et al. (2003). All these papers deal with lytic infections and mass
action incidence functions (Nold, 1980; Dietz and Schenzle, 1985; McCallum
et al., 2001).

In this paper, we focus on modelling the influence of lysogenic infections
and proportionate mixing incidence function (frequency-dependent transmission) on the local and spatio-temporal dynamics of interacting phytoplankton
and zooplankton. Furthermore, the impact of multiplicative noise (Allen, 2003;
Anishenko et al., 2003) is investigated.

4.2   The mathematical model

The model by Scheffer (1991a) for the prey-predator dynamics of phytoplankton
P and zooplankton Z is used as the starting point. It reads in time t and two
spatial dimensions $\vec{r} = \{x, y\}$ with dimensionless quantities, scaled following
Pascual (1993)

$$\frac{\partial P}{\partial t} = rP(1 - P) - \frac{aP}{1 + bP}Z + d\,\Delta P, \qquad (4.1)$$

$$\frac{\partial Z}{\partial t} = \frac{aP}{1 + bP}Z - m_3\,Z + d\,\Delta Z. \qquad (4.2)$$

There is logistic growth of the phytoplankton with intrinsic rate r and Holling-type II grazing with maximum rate a as well as natural mortality with rate m_3 of the zooplankton. The growth rate r is scaled as the ratio of local rate r_{loc} and spatial mean $\langle r \rangle$. The diffusion coefficient d describes eddy diffusion. Therefore, it must be equal for both species. The dynamics of a top predator, i.e., planktivorous fish is neglected because the focus of this paper is on the influence of virally infected phytoplankton. The phytoplankton population P is split into a susceptible part X_1 and an infected portion X_2. Zooplankton is simply renamed to X_3. Then, the model system reads for symmetric inter- and intraspecific competition of susceptibles and infected

$$\frac{\partial X_i(\vec{r},t)}{\partial t} = f_i\left[\mathbf{X}(\vec{r},t)\right] + d\,\Delta X_i(\vec{r},t)\,, \ i = 1,2,3\,; \tag{4.3}$$

where

$$f_1 = r_1 X_1 \left(1 - X_1 - X_2\right) - \frac{a X_1}{1 + b\left(X_1 + X_2\right)} X_3 - \lambda \frac{X_1 X_2}{X_1 + X_2}\,, \tag{4.3a}$$

$$f_2 = r_2 X_2 \left(1 - X_1 - X_2\right) - \frac{a X_2}{1 + b\left(X_1 + X_2\right)} X_3 + \lambda \frac{X_1 X_2}{X_1 + X_2} - m_2 X_2\,, \tag{4.3b}$$

$$f_3 = \frac{a(X_1 + X_2)}{1 + b(X_1 + X_2)} X_3 - m_3 X_3\,. \tag{4.3c}$$

Proportionate mixing with transmission coefficient λ as well as an additional disease-induced mortality of infected (virulence) with rate m_2 are assumed. The vector of population densities is $\mathbf{X} = \{X_1, X_2, X_3\}$. In the case of lytic infection, the first term on the right-hand side of eq. (4.3b) would describe the losses due to natural mortality and competition. Here, lysogenic infections with $r_1 = r_2 = r$ will be considered. However, it is a highly simplified model because the growth rate of susceptibles is usually higher than that of infected (Suttle et al., 1990). Furthermore, the lysogenic replication cycle of viruses is very sensitive to environmental changes and very quickly switches to the lytic cycle. This is left for the second part of the paper.

Furthermore, multiplicative noise is introduced in eqs. (4.3) in order to study environmental fluctuations, i.e.,

$$\frac{\partial X_i(\vec{r},t)}{\partial t} = f_i\left[\mathbf{X}(\vec{r},t)\right] + d\,\Delta X_i(\vec{r},t) + \omega_i\left[\mathbf{X}(\vec{r},t)\right] \cdot \xi_i(\vec{r},t)\,, \ i = 1,2,3\,; \tag{4.4}$$

where $\xi_i(\vec{r},t)$ is a spatiotemporal white Gaussian noise, i.e., a random Gaussian field with zero mean and delta correlation

$$\langle \xi_i(\vec{r},t) \rangle = 0\,, \ \langle \xi_i(\vec{r}_1,t_1)\,\xi_i(\vec{r}_2,t_2) \rangle = \delta(\vec{r}_1 - \vec{r}_2)\,\delta(t_1 - t_2)\,, \ i = 1,2,3\,. \tag{4.4a}$$

$\omega_i\left[\mathbf{X}(\vec{r},t)\right]$ is the density-dependent noise intensity. The stochastic modelling of population dynamics requires this density dependence, i.e., multiplicative noise. Throughout this paper, it is chosen

$$\omega_i\left[\mathbf{X}(\vec{r},t)\right] = \omega X_i(\vec{r},t)\,, \ i = 1,2,3\,; \ \omega = \text{const.} \tag{4.4b}$$

4.3   The deterministic local dynamics

At first, the local dynamics is studied, i.e., we look for stationary and oscillatory solutions of system (4.3) for $d = 0$. Stationary solutions are marked by X_i^{Sno}, $i = 1, 2, 3$. Furthermore, we introduce the parameter

$$m_3^{cr} = \frac{a\left(X_1^{Sno} + X_2^{Sno}\right)}{1 + b\left(X_1^{Sno} + X_2^{Sno}\right)}.$$

Analytical and numerical investigations yield the following selected equilibria:

0) Trivial solution $X_1^{S00} = X_2^{S00} = X_3^{S00} = 0$, always unstable;

1) Endemic states with and without predation:

   a) $X_1^{S11} > 0$, $X_2^{S11} > 0$, $X_3^{S11} > 0$ if $m_2 = \lambda$ and $m_3 < m_3^{cr}$, oscillatory or non-oscillatory stable, multiple stable equilibria possible,

   b) $X_1^{S12} > 0$, $X_2^{S12} > 0$, $X_3^{S12} = 0$ if $m_2 = \lambda$ and $m_3 > m_3^{cr}$, non-oscillatory stable;

2) Extinction of infected with and without predation:

   a) $X_1^{S21} > 0$, $X_2^{S21} = 0$, $X_3^{S21} > 0$ if $m_2 > \lambda$ and $m_3 < m_3^{cr}$, oscillatory or non-oscillatory stable,

   b) $X_1^{S22} > 0$, $X_2^{S22} = 0$, $X_3^{S22} = 0$ if $m_2 > \lambda$ and $m_3 > m_3^{cr}$, non-oscillatory stable,

3) Extinction of susceptibles with and without predation:

   a) $X_1^{S31} = 0$, $X_2^{S31} > 0$, $X_3^{S31} > 0$ if $m_2 < \lambda$ and $m_3 < m_3^{cr}$, oscillatory or non-oscillatory stable,

   b) $X_1^{S32} = 0$, $X_2^{S32} > 0$, $X_3^{S32} = 0$ if $m_2 < \lambda$ and $m_3 > m_3^{cr}$, non-oscillatory stable.

For $m_2 > \lambda$, the infected go extinct (solutions 2a and 2b), for $m_2 < \lambda$, the susceptibles do (solutions 3a and 3b). In the case of $m_2 = \lambda$, susceptibles and infected coexist (endemic states 1a and 1b). Because of the symmetry of the growth terms of susceptibles and infected, the initial conditions determine their final dominance in the endemic state, i.e., if $X_1(t = 0) > X_2(t = 0)$ then $X_1(t) > X_2(t) \, \forall \, t$. A corresponding example is presented in Figs. 4.1 for $r = 1$ and $a = b = 5$. These three parameter values will be kept for all simulations.

The stable prey-predator oscillations are presentations of solutions (2a), (3a) and (1a), respectively.

The local dynamics of model (4.3) is well-known and simple for a single prey. There can be a Hopf bifurcation point, e.g., for decreasing mortality m_3 of the predator. For the parameters in Figs. 4.1, this point has been already passed. A slight increase of m_3 from 0.5 to 0.625 yields multiple equilibria and also demonstrates the dependence of the endemic states on the initial conditions. In Fig. 4.2a for $X_1(0) > X_2(0)$, one still finds the oscillatory solution (1a). However, the opposite choice $X_2(0) > X_1(0)$ in Fig. 4.2b, results in damping of the oscillations, i.e., the different initial conditions belong to different basins of attraction in phase space.

After this rough investigation of the deterministic local behaviour, we proceed now to the stochastic spatial dynamics.

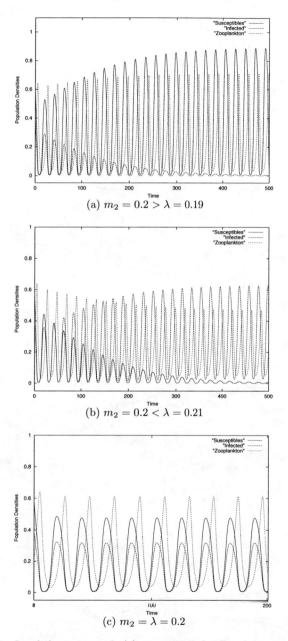

(a) $m_2 = 0.2 > \lambda = 0.19$

(b) $m_2 = 0.2 < \lambda = 0.21$

(c) $m_2 = \lambda = 0.2$

Figure 4.1: Local dynamics with (a) extinction of infected for $m_2 > \lambda$, (b) extinction of susceptibles for $m_2 < \lambda$ and (c) coexistence of susceptibles X_1, infected X_2 and zooplankton X_3 for $m_2 = \lambda$, $m_3 = 0.5$.

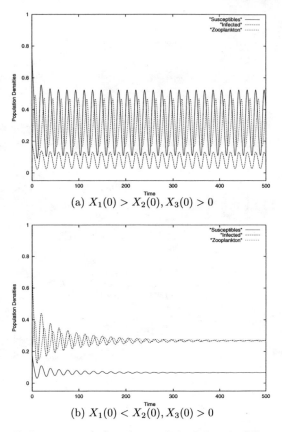

(a) $X_1(0) > X_2(0), X_3(0) > 0$

(b) $X_1(0) < X_2(0), X_3(0) > 0$

Figure 4.2: Endemic state (1a) with initial conditions in different basins of attraction: (a) stable and (b) damped oscillation. $m_2 = \lambda = 0.2, m_3 = 0.625$.

4.4 The deterministic and stochastic spatial dynamics

Much has been published about the spatiotemporal selforganization in prey-predator communities, modelled by reaction-diffusion(-advection) equations, cf. the references in the introduction. Much less is known about equation-based modelling of the spatial spread of epidemics, a small collection of papers includes Grenfell et al. (2001); Abramson et al. (2003) and Zhdanov (2003).

In this paper, we consider the spatiotemporal dynamics of the plankton model (4.4), i.e., zooplankton, grazing on susceptible and virally infected phytoplankton, under the influence of environmental noise and diffusing in horizontally two-dimensional space. The diffusion terms have been integrated using the semi-implicit Peaceman-Rachford alternating direction scheme, cf. Thomas (1995). For the interactions and the Stratonovich integral of the noise terms, the explicit Euler-Maruyama scheme has been applied (Kloeden and Platen, 1992; Higham, 2001).

The following series of figures summarizes the results of the spatiotemporal simulations for growth and interaction parameters from section 4.3, but now including diffusion and noise.

Periodic boundary conditions have been chosen for all simulations.

The initial conditions are localized patches in empty space, and they are the same for deterministic and stochastic simulations. They can be seen in the left column of all following figures. The first two rows show the dynamics of the susceptibles for deterministic and stochastic conditions, the two middle rows show the infected and the two lower rows the zooplankton.

For Figs. 4.3, 4.4 and 4.5, there are two patches, one with zooplankton surrounded by susceptible phytoplankton (upper part of the model area) and one with zooplankton surrounded by infected (on the right of the model area). For Fig. 4.6 and 4.7, there are central patches of all three species. In Fig. 4.6, susceptibles are ahead of infected that are ahead of zooplankton. In Fig. 4.7, infected are ahead of susceptibles that are ahead of zooplankton. In all figures, this special initial configuration leads at first to the propagation of concentric waves for the deterministic case in rows 1, 3 and 5. For the stochastic case in rows 2, 4 and 6, these (naturally unrealistic) waves are immediately blurred and only a leading diffusive front remains.

In Figs. 4.3, one can see the final spatial coexistence of all three species for $m_2 = \lambda$. The localized initial patches generate concentric waves that break up after collision and form spiral waves in a deterministic environment. The noise only blurs these unrealistic patterns. The grey scale changes from high population densities in white colour to vanishing densities in black.

This changes for $m_2 > \lambda$ and $m_2 < \lambda$ in Figs. 4.4 and 4.5, respectively. Whereas in the deterministic case infected or susceptibles go extinct, the noise enhances their survival and spread under unfavourable conditions.

In Figs. 4.6, the deterministic simulations yield the dynamic stabilization of the locally unstable focus in space and a long plateau is formed with a leading diffusive front ahead, cf. Petrovskii and Malchow (2000); Malchow and Petrovskii (2002). Furthermore, the infected are somehow trapped in the center and go almost extinct. The noise enhances the "escape", spread and survival of the infected.

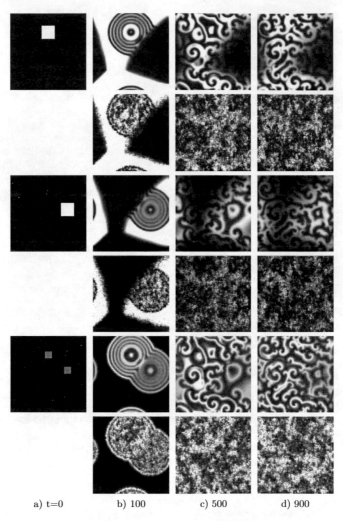

a) t=0   b) 100   c) 500   d) 900

Figure 4.3: Spatial coexistence of susceptibles (two upper rows), infected (two middle rows) and zooplankton (two lower rows) for $m_2 = \lambda = 0.2$, $m_3 = 0.5$, $d = 0.05$. No noise $\omega = 0$ and 0.25 noise intensity, respectively, with equal initial conditions (left column). Periodic boundary conditions.

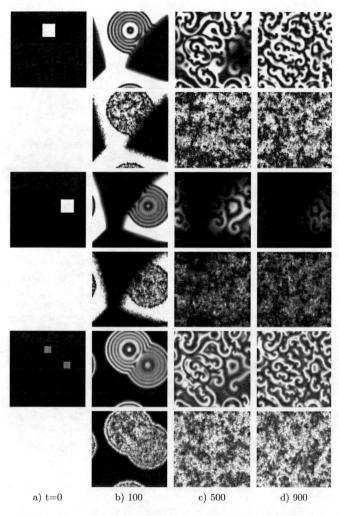

a) t=0   b) 100   c) 500   d) 900

Figure 4.4: Spatial coexistence of susceptibles (two upper rows) and zooplankton (two lower rows). Extinction of infected (third row) for $m_2 = 0.2 > \lambda = 0.19$, $m_3 = 0.5$, $d = 0.05$ and no noise. Survival of infected for $\omega = 0.25$ noise intensity (fourth row).

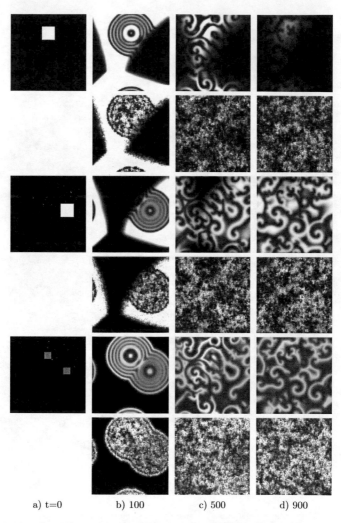

a) t=0   b) 100   c) 500   d) 900

Figure 4.5: Spatial coexistence of infected (two middle rows) and zooplankton (two lower rows). Extinction of susceptibles (first row) for $m_2 = 0.2 < \lambda = 0.21$, $m_3 = 0.5$, $d = 0.05$ and no noise. Survival of susceptibles for $\omega = 0.25$ noise intensity (second row).

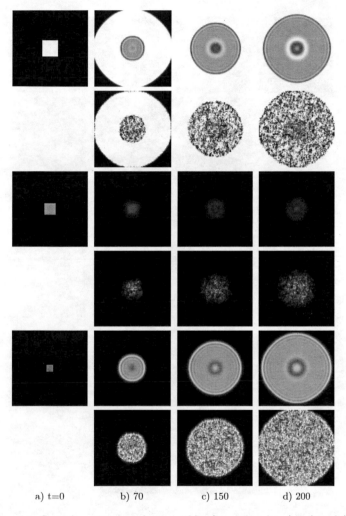

Figure 4.6: Spatial coexistence of susceptibles (two upper rows), infected (two middle rows) and zooplankton (two lower rows) for $m_2 = \lambda = 0.2$, $m_3 = 0.625$, $d = 0.05$. Without noise trapping and almost extinction of infected in the center (third row). With $\omega = 0.25$ noise intensity noise-enhanced survival and escape of infected (fourth row). Phenomenon of dynamic stabilization of a locally unstable equilibrium (first and fifth row).

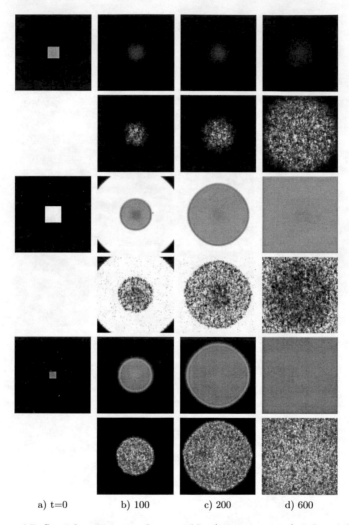

a) t=0   b) 100   c) 200   d) 600

Figure 4.7: Spatial coexistence of susceptibles (two upper rows), infected (two middle rows) and zooplankton (two lower rows) for $m_2 = \lambda = 0.2$, $m_3 = 0.625$, $d = 0.05$. Without noise trapping and almost extinction of suceptibles in the center (first row). With $\omega = 0.25$ noise intensity noise-enhanced survival and escape of susceptibles (second row).

In Figs. 4.7, the dynamic stabilization is not so clearly seen. However, the noise enhances the "escape", spread and survival of the susceptibles here.

4.5 Conclusions

A conceptual biomass-based model of phytoplankton-zooplankton prey-predator dynamics has been investigated for temporal, spatial and spatio-temporal dissipative pattern formation in a deterministic and noisy environment, respectively. It has been assumed that the phytoplankton is partly virally infected and the virus has a lysogenic replication cycle, i.e., also the infected phytoplankton is still able to reproduce.

The equal growth rates of susceptibles and infected have led to the situation that, in a non-fluctuating environment, the ratio of the mortality of the infected and the transmission rate of the infection controls coexistence, survival or extinction of susceptibles and infected. A fluctuating environment enhances the survival and the spatial spread of the "endangered" species. However, noise has not only supported the spatiotemporal coexistence of susceptibles and infected but it has been necessary to blur distinct artificial population structures like target patterns or spirals and to generate more realistic fuzzy patterns.

Forthcoming work has to consider differing growth rates of susceptible and infected species as well as the critical noise-induced switch from lysogenic to lytic viral replications and the resulting spatiotemporal dynamics of the plankton populations.

Acknowledgement

H.M. is thankful to Kay D. Bidle (IMCS, Rutgers University) for some helpful advice on lysogenic and lytic viral infections of phytoplankton as well as for references to relevant publications.

This work has been partially supported by Deutsche Forschungsgemeinschaft, grant no. 436 RUS 113/631.

References

Abramson, G., Kenkre, V. M., Yates, T. L., Parmenter, R. R. (2003). Traveling waves of infection in the hantavirus epidemics. *Bulletin of Mathematical Biology* 65, 519–534.

Allen, L. J. S. (ed.) (2003). *An introduction to stochastic processes with applications to biology*. Pearson Education, Upper Saddle River NJ.

Anishenko, V. S., Astakov, V., Neiman, A. B., Vadivasova, T. E., Schimansky-Geier, L. (2003). *Nonlinear dynamics of chaotic and stochastic systems. Tutorial and modern developments*. Springer Series in Synergetics, Springer, Berlin.

Beltrami, E., Carroll, T. O. (1994). Modelling the role of viral disease in recurrent phytoplankton blooms. *Journal of Mathematical Biology* 32, 857–863.

Chattopadhyay, J., Pal, S. (2002). Viral infection on phytoplankton-zooplankton system – a mathematical model. *Ecological Modelling* 151, 15–28.

Chattopadhyay, J., Sarkar, R. R., Pal, S. (2003). Dynamics of nutrient-phytoplankton interaction in the presence of viral infection. *BioSystems* 68, 5–17.

Dietz, K., Schenzle, D. (1985). Proportionate mixing models for age-dependent infection transmission. *Journal of Mathematical Biology* 22, 117–120.

Fuhrman, J. A. (1999). Marine viruses and their biogeochemical and ecological effects. *Nature* 399, 541–548.

Grenfell, B. T., Bjørnstad, O. N., Kappey, J. (2001). Travelling waves and spatial hierarchies in measles epidemics. *Nature* 414, 716–723.

Higham, D. (2001). An algorithmic introduction to numerical simulation of stochastic differential equations. *SIAM Review* 43(3), 525–546.

Jacquet, S., Heldal, M., Iglesias-Rodriguez, D., Larsen, A., Wilson, W., Bratbak, G. (2002). Flow cytometric analysis of an *Emiliana huxleyi* bloom terminated by viral infection. *Aquatic Microbial Ecology* 27, 111–124.

Jiang, S. C., Paul, J. H. (1998). Significance of lysogeny in the marine environment: studies with isolates and a model of lysogenic phage production. *Microbial Ecology* 35(3), 235–243.

Kloeden, P. E., Platen, E. (1992). *Numerical solution of stochastic differential equations*. Springer, Berlin.

Kuznetsov, Y. A., Muratori, S., Rinaldi, S. (1992). Bifurcations and chaos in a periodic predator-prey model. *International Journal of Bifurcation and Chaos* 2, 117–128.

Malchow, H. (1993). Spatio-temporal pattern formation in nonlinear nonequilibrium plankton dynamics. *Proceedings of the Royal Society of London B* 251, 103–109.

Malchow, H. (1996). Nonlinear plankton dynamics and pattern formation in an ecohydrodynamic model system. *Journal of Marine Systems* 7(2-4), 193–202.

Malchow, H. (2000a). Motional instabilities in predator-prey systems. *Journal of Theoretical Biology* 204, 639–647.

Malchow, H. (2000b). Nonequilibrium spatio-temporal patterns in models of nonlinear plankton dynamics. *Freshwater Biology* 45, 239–251.

Malchow, H., Hilker, F. M., Petrovskii, S. V. (2004a). Noise and productivity dependence of spatiotemporal pattern formation in a prey-predator system. *Discrete and Continuous Dynamical Systems B* 4(3), 705–711.

Malchow, H., Medvinsky, A. B., Petrovskii, S. V. (2004b). Patterns in models of plankton dynamics in a heterogeneous environment. In *Handbook of scaling methods in aquatic ecology: measurement, analysis, simulation* (Seuront, L., Strutton, P. G., eds.). CRC Press, Boca Raton, pp. 401–410.

Malchow, H., Petrovskii, S. V. (2002). Dynamical stabilization of an unstable equilibrium in chemical and biological systems. *Mathematical and Computer Modelling* 36, 307–319.

Malchow, H., Petrovskii, S. V., Hilker, F. M. (2003). Model of spatiotemporal pattern formation in plankton dynamics. *Nova Acta Leopoldina NF* 88(332), 325–340.

Malchow, H., Petrovskii, S. V., Medvinsky, A. B. (2001). Pattern formation in models of plankton dynamics. A synthesis. *Oceanologica Acta* 24(5), 479–487.

Malchow, H., Petrovskii, S. V., Medvinsky, A. B. (2002). Numerical study of plankton-fish dynamics in a spatially structured and noisy environment. *Ecological Modelling* 149, 247–255.

McCallum, H., Barlow, N., Hone, J. (2001). How should pathogen transmission be modelled? *Trends in Ecology & Evolution* 16(6), 295–300.

McDaniel, L., Houchin, L. A., Williamson, S. J., Paul, J. H. (2002). Lysogeny in *Synechococcus*. *Nature* 415, 496.

Medvinsky, A. B., Petrovskii, S. V., Tikhonova, I. A., Malchow, H., Li, B.-L. (2002). Spatiotemporal complexity of plankton and fish dynamics. *SIAM Review* 44(3), 311–370.

Medvinsky, A. B., Tikhonov, D. A., Enderlein, J., Malchow, H. (2000). Fish and plankton interplay determines both plankton spatio-temporal pattern formation and fish school walks. A theoretical study. *Nonlinear Dynamics, Psychology and Life Sciences* 4(2), 135–152.

Menzinger, M., Rovinsky, A. B. (1995). The differential flow instabilities. In *Chemical waves and patterns* (Kapral, R., Showalter, K., eds.). No. 10 in Understanding Chemical Reactivity, Kluwer, Dordrecht, pp. 365–397.

Nold, A. (1980). Heterogeneity in disease-transmission modeling. *Mathematical Biosciences* 52, 227–2402.

Ortmann, A. C., Lawrence, J. E., Suttle, C. A. (2002). Lysogeny and lytic viral production during a bloom of the cyanobacterium *Synechococcus* spp. *Microbial Ecology* 43, 225–231.

Pascual, M. (1993). Diffusion-induced chaos in a spatial predator-prey system. *Proceedings of the Royal Society of London B* 251, 1–7.

Petrovskii, S. V., Malchow, H. (1999). A minimal model of pattern formation in a prey-predator system. *Mathematical and Computer Modelling* 29, 49–63.

Petrovskii, S. V., Malchow, H. (2000). Critical phenomena in plankton communities: KISS model revisited. *Nonlinear Analysis: Real World Applications* 1, 37–51.

Petrovskii, S. V., Malchow, H. (2001). Wave of chaos: new mechanism of pattern formation in spatio-temporal population dynamics. *Theoretical Population Biology* 59(2), 157–174.

Rinaldi, S., Muratori, S., Kuznetsov, Y. (1993). Multiple attractors, catastrophes and chaos in seasonally perturbed predator-prey communities. *Bulletin of Mathematical Biology* 55, 15–35.

Sarkar, R. R., Chattopadhayay, J. (2003). Occurence of planktonic blooms under environmental fluctuations and its possible control mechanism – mathematical models and experimental observations. *Journal of Theoretical Biology* 224, 501–516.

Satnoianu, R. A., Menzinger, M. (2000). Non-Turing stationary patterns in flow-distributed oscillators with general diffusion and flow rates. *Physical Review E* 62(1), 113–119.

Satnoianu, R. A., Menzinger, M., Maini, P. K. (2000). Turing instabilities in general systems. *Journal of Mathematical Biology* 41, 493–512.

Scheffer, M. (1991a). Fish and nutrients interplay determines algal biomass: a minimal model. *Oikos* 62, 271–282.

Scheffer, M. (1991b). Should we expect strange attractors behind plankton dynamics - and if so, should we bother? *Journal of Plankton Research* 13, 1291–1305.

Scheffer, M., Rinaldi, S., Kuznetsov, Y. A., van Nes, E. H. (1997). Seasonal dynamics of daphnia and algae explained as a periodically forced predator-prey system. *Oikos* 80, 519–532.

Sherratt, J. A., Lewis, M. A., Fowler, A. C. (1995). Ecological chaos in the wake of invasion. *Proceedings of the National Academy of Sciences of the United States of America* 92, 2524–2528.

Steele, J., Henderson, E. W. (1992). A simple model for plankton patchiness. *Journal of Plankton Research* 14, 1397–1403.

Steffen, E., Malchow, H., Medvinsky, A. B. (1997). Effects of seasonal perturbation on a model plankton community. *Environmental Modeling and Assessment* 2, 43–48.

Suttle, C. A. (2000). Ecological, evolutionary, and geochemical consequences of viral infection of cyanobacteria and eukaryotic algae. In *Viral ecology* (Hurst, C. J., ed.). Academic Press, San Diego, pp. 247–296.

Suttle, C. A., Chan, A. M., Cottrell, M. T. (1990). Infection of phytoplankton by viruses and reduction of primary productivity. *Nature* 347, 467–469.

Thomas, J. W. (1995). *Numerical partial differential equations: Finite difference methods*. Springer-Verlag, New York.

Truscott, J. E., Brindley, J. (1994). Ocean plankton populations as excitable media. *Bulletin of Mathematical Biology* 56, 981–998.

Wilcox, R., Fuhrman, J. (1994). Bacterial viruses in coastal seawater: lytic rather than lysogenic production. *Marine Ecology Progress Series* 114, 35–45.

Wommack, K. E., Colwell, R. R. (2000). Virioplankton: viruses in aquatic ecosystems. *Microbiology and Molecular Biology Reviews* 64(1), 69–114.

Zhdanov, V. P. (2003). Propagation of infection and the predator-prey interplay. *Journal of Theoretical Biology* 225, 489–492.

Chapter 5

Spatiotemporal patterns in an excitable plankton system with lysogenic viral infection

Horst Malchow, Frank M. Hilker, Ram Rup Sarkar and Klaus Brauer

Institute of Environmental Systems Research, Department of Mathematics and Computer Science, University of Osnabrück, 49069 Osnabrück, Germany

To appear in *Mathematical and Computer Modelling*.[1]
Accepted October 11, 2004.

Abstract

An excitable model of phytoplankton-zooplankton dynamics is considered for the case of lysogenic viral infection of the phytoplankton population. The phytoplankton population is split into a susceptible (S) and an infected (I) part. Both parts grow logistically, limited by a common carrying capacity. Zooplankton (Z) is grazing on susceptibles and infected, following a Holling-type III functional response. The local analysis of the S-I-Z differential equations yields a number of stationary and/or oscillatory regimes and their combinations. Correspondingly interesting is the spatiotemporal behaviour, modelled by stochastic reaction-diffusion equations. Spatial spread or suppression of infection will be presented just as well as competition of concentric and/or spiral population waves for space. The external noise can enhance the survival and spread of susceptibles and infected, respectively, that would go extinct in a deterministic environment. In the parameter range of excitability, noise can induce local blooms of susceptibles and infected.

---

[1]Printed with kind permission from Elsevier.

5.1   Introduction

Viruses are evidently the most abundant entities in the sea and the question may arise whether they control ocean life. However, there is much less known about marine viruses and their role in aquatic ecosystems and the species that they infect, than about plankton patchiness and blooming, for reviews, cf. Fuhrman (1999). A number of studies (Bergh et al., 1989; Suttle et al., 1990; Wilhelm and Suttle, 1999; Tarutani et al., 2000; Wommack and Colwell, 2000) shows the presence of pathogenic viruses in phytoplankton communities. Fuhrman (1999) has reviewed the nature of marine viruses and their ecological as well as bio-geological effects. Suttle et al. (1990) have shown by using electron microscopy that the viral disease can infect bacteria and phytoplankton in coastal water. Parasites may modify the behaviour of the infected members of the prey population. Virus-like particles are described for many eukaryotic algae (van Etten et al., 1991; Reiser, 1993), cyanobacteria (Suttle et al., 1993) and natural phytoplankton communities (Peduzzi and Weinbauer, 1993). There is some evidence that viral infection might accelerate the termination of phytoplankton blooms (Jacquet et al., 2002; Gastrich et al., 2004). Viruses are held responsible for the collapse of *Emiliania huxleyi* blooms in mesocosms (Bratbak et al., 1995) and in the North Sea (Brussard et al., 1996) and are shown to induce lysis of *Chrysochromulina* (Suttle and Chan, 1993). Because most viruses are strain-specific, they can increase genetic diversity (Nagasaki and Yamaguchi, 1997). Nevertheless, despite the increasing number of reports, the role of viral infection in the phytoplankton population is still far from understood.

Viral infections of phytoplankton cells can be lysogenic or lytic. The understanding of the importance of lysogeny is just at the beginning (Wilcox and Fuhrman, 1994; Jiang and Paul, 1998; McDaniel et al., 2002; Ortmann et al., 2002). Contrary to lytic infections with destruction and without reproduction of the host cell, lysogenic infections are a strategy whereby viruses integrate their genome into the host's genome. As the host reproduces and duplicates its genome, the viral genome reproduces, too.

Mathematical models of the dynamics of virally infected phytoplankton populations are rare as well, the already classical publication is by Beltrami and Carroll (1994). More recent work is of Chattopadhyay and Pal (2002) and Chattopadhyay et al. (2003). The latter deal with lytic infections and mass action incidence functions. Malchow et al. (2004b) observed oscillations and waves in a phytoplankton-zooplankton system with Holling-type II grazing under lysogenic viral infection and frequency-dependent transmission.

Numerous papers have been published about pattern formation and chaos in minimal prey-predator models of phytoplankton-zooplankton dynamics (Scheffer, 1991a; Malchow, 1993; Pascual, 1993; Truscott and Brindley, 1994; Malchow, 1996, 2000b; Malchow et al., 2001, 2004c). Different routes to local and spatiotemporal chaos (Scheffer, 1991b; Kuznetsov et al., 1992; Rinaldi et al., 1993; Sherratt et al., 1995; Scheffer et al., 1997; Steffen et al., 1997; Petrovskii and Malchow, 1999, 2001; Malchow et al., 2002), diffusion- and differential-flow-induced standing and travelling waves (Malchow, 1993; Menzinger and Rovinsky, 1995; Malchow, 2000a; Satnoianu and Menzinger, 2000; Satnoianu et al., 2000; Malchow et al., 2003) as well as target patterns and spiral waves (Medvinsky et al., 2000, 2002) have been found. Also the impact of external noise on patchiness and transitions between alternative stable population states has been studied

(Malchow et al., 2002; Steele and Henderson, 1992; Sarkar and Chattopadhayay, 2003; Malchow et al., 2004a).

In this paper, we focus on modelling the influence of lysogenic infections and proportionate mixing incidence function (frequency-dependent transmission) (Nold, 1980; Dietz and Schenzle, 1985; McCallum et al., 2001) on the local and spatio-temporal dynamics of interacting phytoplankton and zooplankton with Holling-type III grazing, i.e., with excitable dynamics. The latter has been introduced by Truscott and Brindley (1994) to model recurrent phytoplankton blooms. Furthermore, the impact of multiplicative noise (Allen, 2003; Anishenko et al., 2003) is investigated.

5.2 The mathematical model

The Truscott-Brindley model (1994) for the prey-predator dynamics of phytoplankton P and zooplankton Z at time t and location $\vec{r} = \{x, y\}$ reads in dimensionless quantities

$$\frac{\partial P}{\partial t} = rP\left(1 - P\right) - \frac{a^2 P^2}{1 + b^2 P^2} Z + d\,\Delta P, \qquad (5.1)$$

$$\frac{\partial Z}{\partial t} = \frac{a^2 P^2}{1 + b^2 P^2} Z - m_3 Z + d\,\Delta Z. \qquad (5.2)$$

There is logistic growth of the phytoplankton with intrinsic rate r and Holling-type III grazing with maximum rate a^2/b^2 as well as natural mortality of zooplankton with rate m_3. The growth rate r is scaled as the ratio of local rate r_{loc} and spatial mean $\langle r \rangle$. The diffusion coefficient d describes eddy diffusion. Therefore, it must be equal for both species. The effects of nutrient supply and planktivorous fish are neglected because the focus of this paper is on the influence of virally infected phytoplankton. The phytoplankton population P is split into a susceptible part X_1 and an infected portion X_2. Zooplankton is simply renamed to X_3. Then, the model system reads for symmetric inter- and intraspecific competition of susceptibles and infected

$$\frac{\partial X_i(\vec{r}, t)}{\partial t} = f_i\left[\mathbf{X}(\vec{r}, t)\right] + d\,\Delta X_i(\vec{r}, t), \; i = 1, 2, 3; \qquad (5.3)$$

where

$$f_1 = r_1 X_1\left(1 - X_1 - X_2\right) - \frac{a^2 X_1\left(X_1 + X_2\right)}{1 + b^2\left(X_1 + X_2\right)^2} X_3 - \lambda \frac{X_1 X_2}{X_1 + X_2}, \qquad (5.3a)$$

$$f_2 = r_2 X_2\left(1 - X_1 - X_2\right) - \frac{a^2 X_2\left(X_1 + X_2\right)}{1 + b^2\left(X_1 + X_2\right)^2} X_3 + \lambda \frac{X_1 X_2}{X_1 + X_2} - m_2 X_2, \qquad (5.3b)$$

$$f_3 = \frac{a^2(X_1 + X_2)^2}{1 + b^2(X_1 + X_2)^2} X_3 - m_3 X_3. \qquad (5.3c)$$

Proportionate mixing transmission with rate λ as well as an additional disease-induced mortality of infected (virulence) with rate m_2 are assumed. The vector of population densities is $\mathbf{X} = \{X_1, X_2, X_3\}$. In the case of lytic infection, the first term on the right-hand side of eq. (5.3b) would describe the losses due to natural mortality and competition. Here, lysogenic infections with $r_1 = r_2 = r$ will be considered. The lysogenic replication cycle of viruses is very sensitive to environmental variability and may switch to the lytic cycle. This situation is not considered here.

Furthermore, multiplicative noise is introduced in eqs. (5.3) in order to study environmental fluctuations, i.e.,

$$\frac{\partial X_i(\vec{r}, t)}{\partial t} = f_i\left[\mathbf{X}(\vec{r}, t)\right] + d\, \Delta X_i(\vec{r}, t) + \omega_i\left[\mathbf{X}(\vec{r}, t)\right] \cdot \xi_i(\vec{r}, t)\,, \ i = 1, 2, 3; \quad (5.4)$$

where $\xi_i(\vec{r}, t)$ is a spatiotemporal white Gaussian noise, i.e., a random Gaussian field with zero mean and delta correlation

$$\langle \xi_i(\vec{r}, t) \rangle = 0\,, \ \ \langle \xi_i(\vec{r}_1, t_1)\, \xi_i(\vec{r}_2, t_2) \rangle = \delta(\vec{r}_1 - \vec{r}_2)\, \delta(t_1 - t_2)\,, \ i = 1, 2, 3\,. \quad (5.4a)$$

$\omega_i\left[\mathbf{X}(\vec{r}, t)\right]$ is the density-dependent noise intensity. The axiom of parentness in population dynamics requires this density dependence, i.e., multiplicative noise. Throughout this paper, it is chosen

$$\omega_i\left[\mathbf{X}(\vec{r}, t)\right] = \omega X_i(\vec{r}, t)\,, \ i = 1, 2, 3\,; \ \omega = \text{const.} \quad (5.4b)$$

5.3   The local dynamics

At first, the local dynamics is studied, i.e., it is searched for stationary and oscillatory solutions of system (5.3) for $d = 0$. To do that, system (5.3) is simplified through a convenient transformation, then describing the dynamics of the total phytoplankton population $P = X_1 + X_2$, the prevalence $i = X_2/P$ and zooplankton X_3. With $f(P) = a^2 P^2/(1 + b^2 P^2)$, the model equations read

$$\frac{\mathrm{d}\,P}{\mathrm{d}\,t} = \left[r_1(1 - i) + r_2 i \right](1 - P)\, P - f(P) X_3 - m_2 i P,$$
$$\frac{\mathrm{d}\,i}{\mathrm{d}\,t} = \left[(r_2 - r_1)(1 - P) + (\lambda - m_2) \right](1 - i)\, i, \quad (5.5)$$
$$\frac{\mathrm{d}\,X_3}{\mathrm{d}\,t} = \left[f(P) - m_3 \right] X_3.$$

For different ratios of λ and m_2 there exist different stationary solutions. In order to characterize them, the following additional parameters are introduced:

$$i^S = \frac{X_2^S}{X_1^S + X_2^S} \ \ \text{and} \ \ m_3^{cr} = f(P^S) = \frac{a^2\, (X_1^S + X_2^S)^2}{1 + b^2\, (X_1^S + X_2^S)^2}\,.$$

X_k^S, $k = 1, 2, 3$; are the stationary solutions of system (5.3). Both analytical and numerical investigations yield the following selected equilibria for $r_1 = r_2 = r$:

  0) Trivial solution $X_1^S = X_2^S = X_3^S = 0$, i.e., $P^S = i^S = X_3^S = 0$, always unstable;

  1) Extinction of infected with and without predation:

a) $P^S = X_1^S > 0$, $X_2^S = 0$, i.e., $i^S = 0$, $X_3^S = 0$ if $\lambda < m_2$ and $m_3 > m_3^{cr}$, non-oscillatory stable,

b) $P^S = X_1^S > 0$, $X_2^S = 0$, i.e., $i^S = 0$, $X_3^S > 0$ if $\lambda < m_2$, non-oscillatory or oscillatory stable depending on $m_3 > m_3^{cr}$ or $m_3 < m_3^{cr}$, respectively,

2) Extinction of susceptibles with and without predation:

a) $X_1^S = 0$, $P^S = X_2^S > 0$, i.e., $i^S = 1$, $X_3^S = 0$ if $\lambda > m_2$ and $m_3 > m_3^{cr}$, non-oscillatory stable,

b) $X_1^S = 0$, $P^S = X_2^S > 0$, i.e., $i^S = 1$, $X_3^S > 0$ if $\lambda > m_2$, non-oscillatory or oscillatory stable depending on $m_3 > m_3^{cr}$ or $m_3 < m_3^{cr}$, respectively,

3) Endemic states with and without predation:

a) $X_1^S > 0$, $X_2^S > 0$, i.e., $P^S > 0$, $0 < i^S = i(0) = \text{const.} < 1$, $X_3^S = 0$ if $\lambda = m_2$ and $m_3 > m_3^{cr}$, non-oscillatory stable,

b) $X_1^S > 0$, $X_2^S > 0$, i.e., $P^S > 0$, $0 < i^S = i(0) = \text{const.} < 1$, $X_3^S > 0$ if $\lambda = m_2$, non-oscillatory or oscillatory stable depending on $m_3 > m_3^{cr}$ or $m_3 < m_3^{cr}$, respectively.

For $\lambda < m_2$, the infected go extinct (solutions 1a and 1b) and the prevalence i reaches zero. For $\lambda > m_2$, the susceptibles die out (solutions 2a and 2b) and the prevalence approaches unity. In the case of $\lambda = m_2$, susceptibles and infected coexist (endemic states 3a and 3b) and the prevalence remains constant at its initial value. Moreover, if m_3 is greater or less than m_3^{cr}, the system becomes non-oscillatory or oscillatory stable, respectively. For low values of m_3, we observe excitation and the following relaxation to the non-oscillatory stable situation. A corresponding example is presented in the left column of Figs. 5.1 for $r = r_1 = r_2 = 1$ and $a = 3.75, b = 10$. These three parameter values will be kept for all simulations. In the excitable parameter range with weak external noise, we also observe recurrent outbreaks related to planktonic blooming. This is shown in the right column. These stochastic sample runs have qualitatively the same outcome as the deterministic computations. However, one should have in mind that the latter must only hold for the average of a sufficient number of runs. In a noisy environment, there are only certain probabilities for the survival or extinction of the populations.

A slight increase of m_3 yields loss of excitability but oscillations in the system. These are drawn in Figs. 5.2. The dynamics of the prevalence remains unchanged what is a very convenient property of system (5.5).

One can see that the transformation of the local part of model (5.3) to system (5.5) with $r_1 = r_2 = r$ reduces the considerations of deterministic stationarity and stability to a pseudo-two-dimensional problem because the prevalence can take only three values, i.e., zero for $\lambda < m_2$, unity for $\lambda > m_2$ or its initial value for $\lambda = m_2$. This simplifies the computations remarkably. However, for the investigation of the spatiotemporal system with external noise, we proceed with model (5.3).

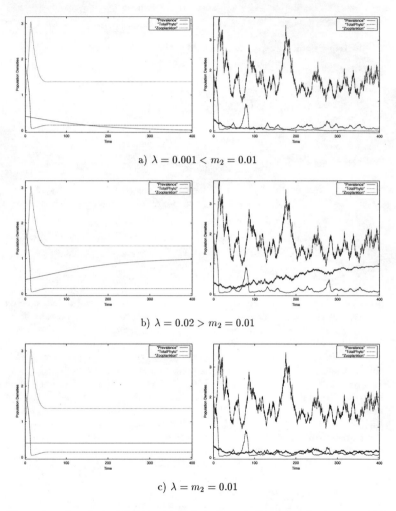

a) $\lambda = 0.001 < m_2 = 0.01$

b) $\lambda = 0.02 > m_2 = 0.01$

c) $\lambda = m_2 = 0.01$

Figure 5.1: Local excitability with (a) decline of prevalence for $\lambda < m_2$, (b) saturation of prevalence for $\lambda > m_2$ and (c) constant prevalence for $\lambda = m_2$, $m_3 = 0.05$. No noise in left column. Localized outbreaks in a noisy environment with $\omega = 0.075$, right column.

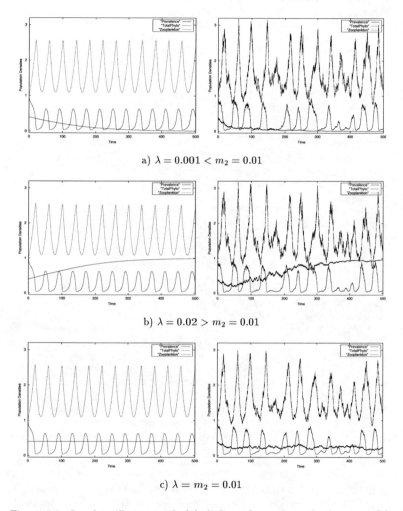

a) $\lambda = 0.001 < m_2 = 0.01$

b) $\lambda = 0.02 > m_2 = 0.01$

c) $\lambda = m_2 = 0.01$

Figure 5.2: Local oscillations with (a) decline of prevalence for $\lambda < m_2$, (b) saturation of prevalence for $\lambda > m_2$, and (c) constant prevalence for $\lambda = m_2$, $m_3 = 0.09$. No noise in left column. Noisy oscillations with $\omega = 0.075$, right column.

5.4   The spatial dynamics

Much has been published about the spatiotemporal selforganization in prey-predator communities, modelled by reaction-diffusion(-advection) equations, cf. the references in the introduction. Much less is known about equation-based modelling of the spatial spread of epidemics, a small collection of papers includes Grenfell et al. (2001); Abramson et al. (2003) and Zhdanov (2003).

In this section, we consider the spatiotemporal dynamics of the plankton model (5.4), i.e., zooplankton, grazing on susceptible and virally infected phytoplankton, under the influence of environmental noise and diffusing in horizontally two-dimensional space. The diffusion terms have been integrated using the semi-implicit Peaceman-Rachford alternating direction scheme, cf. Thomas (1995). For the interactions and the Stratonovich integral of the noise terms, the explicit Euler-Maruyama scheme has been applied (Kloeden and Platen, 1999; Higham, 2001).

The following series of figures summarizes the results of the spatiotemporal simulations for growth and interaction parameters from section 5.3, but now including diffusion and noise. Periodic boundary conditions have been chosen for all simulations.

The initial conditions are localized patches in empty space, and they are the same for deterministic and stochastic simulations. They can be seen in the left column of all following figures. The first two rows show the dynamics of the susceptibles for deterministic and stochastic conditions, the two middle rows show the infected and the two lower rows the zooplankton.

In Figs. 5.3 and 5.4, there are initially two patches, one with zooplankton surrounded by susceptible phytoplankton (upper part of the model area) and one with zooplankton surrounded by infected (on the right of the model area). For Figs. 5.5, there are central and concentric patches of all three species.

In Figs. 5.3, one can see the final spatial coexistence of all three species for $\lambda = m_2$. The localized initial patches generate concentric waves that break up after collision and form spiral waves in a deterministic environment. The noise only blurs these unrealistic patterns. The grey scale changes from high population densities in black colour to vanishing densities in white.

This changes for $\lambda \neq m_2$. Whereas in the deterministic case infected or susceptibles go extinct, respectively, the noise enhances their survival and spread under unfavourable conditions. An example is given in Figs. 5.4 for $\lambda > m_2$, i.e., the deterministic extinction and noise-induced survival and spread of susceptibles. An example for the opposite case is omitted here.

In Figs. 5.5, the deterministic conditions allow for the excitable, non-oscillatory coexistence of susceptibles and infected. Susceptibles are initially ahead of infected that are ahead of zooplankton. This special initial configuration leads to the propagation of diffusive fronts in rows 1, 3 and 5. However, the infected are somehow trapped in the centre and go almost extinct. For the stochastic case in rows 2, 4 and 6, the noise enhances the "escape", spread and survival of the infected. The opposite case is also possible if the infected are initially ahead of susceptibles.

In Figs. 5.6, spatial snapshots of all the three populations at $t = 550$ are provided, without noise in the left column and with noise in the right one. The noise-induced enhancement of the spread of infected is just as readily seen as the localized noise-induced outbreaks.

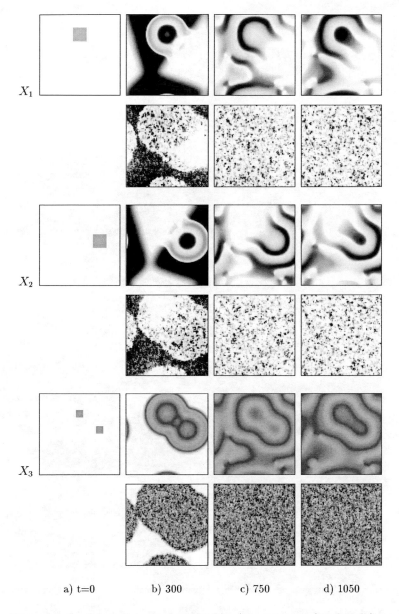

a) t=0   b) 300   c) 750   d) 1050

Figure 5.3: Spatial coexistence of susceptibles (two upper rows), infected (two middle rows) and zooplankton (two lower rows) for $m_2 = \lambda = 0.01$, $m_3 = 0.09$. No noise $\omega = 0$ and 0.25 noise intensity, respectively, with equal initial conditions (left column). Periodic boundary conditions.

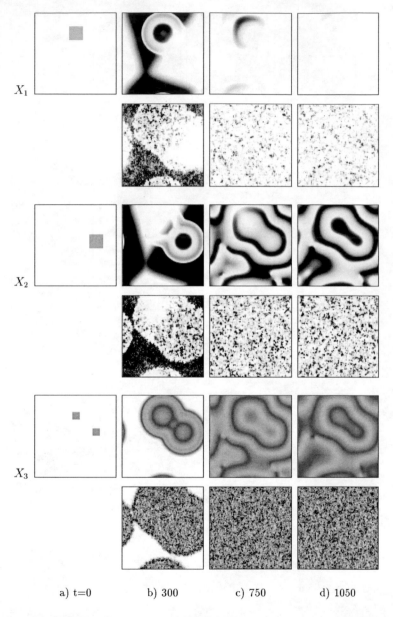

a) t=0   b) 300   c) 750   d) 1050

Figure 5.4: Spatial coexistence of infected (two middle rows) and zooplankton (two lower rows). Extinction of susceptibles (first row) for $m_2 = 0.01 < \lambda = 0.03$, $m_3 = 0.09$, and no noise. Very low survival of susceptibles for $\omega = 0.25$ noise intensity (second row).

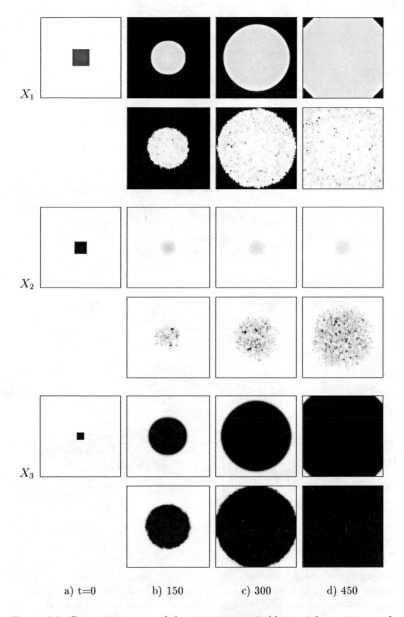

a) t=0   b) 150   c) 300   d) 450

Figure 5.5: Parameter range of deterministic excitable spatial coexistence of susceptibles, infected and zooplankton for $m_2 = \lambda = 0.01$, $m_3 = 0.05$. Without noise trapping and almost extinction of infected in the center (third row). With $\omega = 0.25$ noise intensity noise-enhanced survival and spread of infected (fourth row).

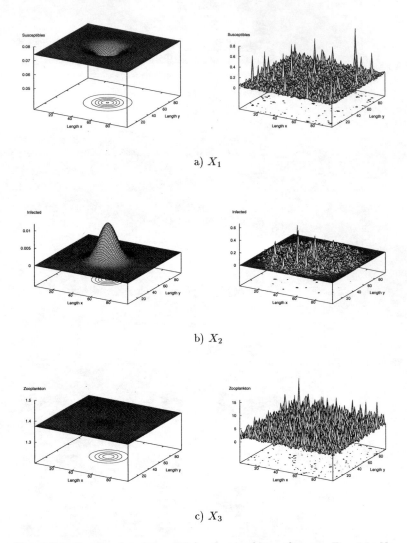

a) X_1

b) X_2

c) X_3

Figure 5.6: Density plot at $t = 550$ for the simulation shown in Figs. 5.5. No noise in left column, noise intensity $\omega = 0.25$ in right column. $m_2 = \lambda = 0.01$, $m_3 = 0.05$. The trapping of infected without noise and their noise-enhanced spread are readily seen. Noise-induced local outbreaks due to the excitability of the system can also be observed.

5.5 Conclusions

A conceptual biomass-based model of phytoplankton-zooplankton prey-predator dynamics has been investigated for temporal, spatial and spatio-temporal dissipative pattern formation in a deterministic and noisy environment, respectively. It has been assumed that the phytoplankton is partly virally infected and the virus has a lysogenic replication cycle, i.e., also the infected phytoplankton is still able to reproduce. Holling-type III zooplankton grazing has been considered in order to study the interplay of excitability, infection and noise.

The equal growth rates of susceptibles and infected have led to the situation that, in a non-fluctuating environment, the ratio of virulence and transmission rate of the infection controls coexistence, survival or extinction of susceptibles and infected, respectively. A fluctuating environment enhances the survival and the spatial spread of the "endangered" species. Furthermore, the noise has induced localized outbreaks or bloom phenomena in the parameter range of excitability. However, noise has not only supported the spatiotemporal coexistence and spread of susceptibles and infected but it has blurred distinct artificial population structures like target patterns or spirals and generated more realistic fuzzy patterns.

Forthcoming work has to include modelling of the transition from lysogenic to lytic viral replications, induced by noise with an intensity above a critical threshold, and its impact on recurrent phytoplankton outbreaks. Furthermore, different incidence functions and noise characteristics and the resulting local and spatiotemporal dynamics of the plankton populations have to be considered.

Acknowledgement

This paper has been completed during a stay of H.M. and F.M.H. in Japan. They are thankful especially to Nanako Shigesada and Hiromi Seno for intensive and fruitful discussions as well as to the Japan Society for the Promotion of Science for the Research and Predoc Fellowships S04716 and PE04533, respectively.

R.R.S. acknowledges a Research Fellowship of the Alexander von Humboldt Foundation, Germany.

References

Abramson, G., Kenkre, V. M., Yates, T. L., Parmenter, R. R. (2003). Traveling waves of infection in the hantavirus epidemics. *Bulletin of Mathematical Biology* 65, 519–534.

Allen, L. J. S. (ed.) (2003). *An introduction to stochastic processes with applications to biology*. Pearson Education, Upper Saddle River NJ.

Anishenko, V. S., Astakov, V., Neiman, A. B., Vadivasova, T. E., Schimansky-Geier, L. (2003). *Nonlinear dynamics of chaotic and stochastic systems. Tutorial and modern developments*. Springer Series in Synergetics, Springer, Berlin.

Beltrami, E., Carroll, T. O. (1994). Modelling the role of viral disease in recurrent phytoplankton blooms. *Journal of Mathematical Biology* 32, 857–863.

Bergh, O., Borsheim, K. Y., Bratbak, G., Heldal, M. (1989). High abundance of viruses found in aquatic environments. *Nature* 340, 467–468.

Bratbak, G., Levasseur, M., Michand, S., Cantin, G., Fernandez, E., Heldal, M. (1995). Viral activity in relation to *Emiliania huxleyi* blooms: a mechanism of DMSP release? *Marine Ecology Progress Series* 128, 133–142.

Brussard, C. P. D., Kempers, R. S., Kop, A. J., Riegman, R., Heldal, M. (1996). Virus-like particles in a summer bloom of *Emiliania huxleyi* in the North Sea. *Aquatic Microbial Ecology* 10(2), 105–113.

Chattopadhyay, J., Pal, S. (2002). Viral infection on phytoplankton-zooplankton system – a mathematical model. *Ecological Modelling* 151, 15–28.

Chattopadhyay, J., Sarkar, R. R., Pal, S. (2003). Dynamics of nutrient-phytoplankton interaction in the presence of viral infection. *BioSystems* 68, 5–17.

Dietz, K., Schenzle, D. (1985). Proportionate mixing models for age-dependent infection transmission. *Journal of Mathematical Biology* 22, 117–120.

Fuhrman, J. A. (1999). Marine viruses and their biogeochemical and ecological effects. *Nature* 399, 541–548.

Gastrich, M. D., Leigh-Bell, J. A., Gobler, C. J., Anderson, O. R., Wilhelm, S. W., Bryan, M. (2004). Viruses as potential regulators of regional brown tide blooms caused by the alga, *Aureococcus anophagefferens*. *Estuaries* 27(1), 112–119.

Grenfell, B. T., Bjørnstad, O. N., Kappey, J. (2001). Travelling waves and spatial hierarchies in measles epidemics. *Nature* 414, 716–723.

Higham, D. (2001). An algorithmic introduction to numerical simulation of stochastic differential equations. *SIAM Review* 43(3), 525–546.

Jacquet, S., Heldal, M., Iglesias-Rodriguez, D., Larsen, A., Wilson, W., Bratbak, G. (2002). Flow cytometric analysis of an *Emiliana huxleyi* bloom terminated by viral infection. *Aquatic Microbial Ecology* 27, 111–124.

Jiang, S. C., Paul, J. H. (1998). Significance of lysogeny in the marine environment: studies with isolates and a model of lysogenic phage production. *Microbial Ecology* 35(3), 235–243.

Kloeden, P. E., Platen, E. (1999). *Numerical solution of stochastic differential equations*. Springer, Berlin.

Kuznetsov, Y. A., Muratori, S., Rinaldi, S. (1992). Bifurcations and chaos in a periodic predator-prey model. *International Journal of Bifurcation and Chaos* 2, 117–128.

Malchow, H. (1993). Spatio-temporal pattern formation in nonlinear nonequilibrium plankton dynamics. *Proceedings of the Royal Society of London B* 251, 103–109.

Malchow, H. (1996). Nonlinear plankton dynamics and pattern formation in an ecohydrodynamic model system. *Journal of Marine Systems* 7(2-4), 193–202.

Malchow, H. (2000a). Motional instabilities in predator-prey systems. *Journal of Theoretical Biology* 204, 639–647.

Malchow, H. (2000b). Nonequilibrium spatio-temporal patterns in models of nonlinear plankton dynamics. *Freshwater Biology* 45, 239–251.

Malchow, H., Hilker, F. M., Petrovskii, S. V. (2004a). Noise and productivity dependence of spatiotemporal pattern formation in a prey-predator system. *Discrete and Continuous Dynamical Systems B* 4(3), 705–711.

Malchow, H., Hilker, F. M., Petrovskii, S. V., Brauer, K. (2004b). Oscillations and waves in a virally infected plankton system: Part I: The lysogenic stage. *Ecological Complexity* 1(3), 211–223.

Malchow, H., Medvinsky, A. B., Petrovskii, S. V. (2004c). Patterns in models of plankton dynamics in a heterogeneous environment. In *Handbook of scaling methods in aquatic ecology: measurement, analysis, simulation* (Seuront, L., Strutton, P. G., eds.). CRC Press, Boca Raton, pp. 401–410.

Malchow, H., Petrovskii, S. V., Hilker, F. M. (2003). Model of spatiotemporal pattern formation in plankton dynamics. *Nova Acta Leopoldina NF* 88(332), 325–340.

Malchow, H., Petrovskii, S. V., Medvinsky, A. B. (2001). Pattern formation in models of plankton dynamics. A synthesis. *Oceanologica Acta* 24(5), 479–487.

Malchow, H., Petrovskii, S. V., Medvinsky, A. B. (2002). Numerical study of plankton-fish dynamics in a spatially structured and noisy environment. *Ecological Modelling* 149, 247–255.

McCallum, H., Barlow, N., Hone, J. (2001). How should pathogen transmission be modelled? *Trends in Ecology & Evolution* 16(6), 295–300.

McDaniel, L., Houchin, L. A., Williamson, S. J., Paul, J. H. (2002). Lysogeny in *Synechococcus*. *Nature* 415, 496.

Medvinsky, A. B., Petrovskii, S. V., Tikhonova, I. A., Malchow, H., Li, B.-L. (2002). Spatiotemporal complexity of plankton and fish dynamics. *SIAM Review* 44(3), 311–370.

Medvinsky, A. B., Tikhonov, D. A., Enderlein, J., Malchow, H. (2000). Fish and plankton interplay determines both plankton spatio-temporal pattern formation and fish school walks. A theoretical study. *Nonlinear Dynamics, Psychology and Life Sciences* 4(2), 135–152.

Menzinger, M., Rovinsky, A. B. (1995). The differential flow instabilities. In *Chomioal wavoo and pattorno* (Kapral, R., Showalter, K., cds.). No. 10 in Understanding Chemical Reactivity, Kluwer, Dordrecht, pp. 365–397.

Nagasaki, K., Yamaguchi, M. (1997). Isolation of a virus infectious to the harmful bloom causing microalga *Heterosigma akashiwo* (Raphidophyceae). *Aquatic Microbial Ecology* 13, 135–140.

Nold, A. (1980). Heterogeneity in disease-transmission modeling. *Mathematical Biosciences* 52, 227–2402.

Ortmann, A. C., Lawrence, J. E., Suttle, C. A. (2002). Lysogeny and lytic viral production during a bloom of the cyanobacterium *Synechococcus* spp. *Microbial Ecology* 43, 225–231.

Pascual, M. (1993). Diffusion-induced chaos in a spatial predator-prey system. *Proceedings of the Royal Society of London B* 251, 1–7.

Peduzzi, P., Weinbauer, M. G. (1993). The submicron size fraction of sea water containing high numbers of virus particles as bioactive agent in unicellular plankton community successions. *Journal of Plankton Research* 15, 1375–1386.

Petrovskii, S. V., Malchow, H. (1999). A minimal model of pattern formation in a prey-predator system. *Mathematical and Computer Modelling* 29, 49–63.

Petrovskii, S. V., Malchow, H. (2001). Wave of chaos: new mechanism of pattern formation in spatio-temporal population dynamics. *Theoretical Population Biology* 59(2), 157–174.

Reiser, W. (1993). Viruses and virus like particles of freshwater and marine eukaryotic algae – a review. *Archiv für Protistenkunde* 143, 257–265.

Rinaldi, S., Muratori, S., Kuznetsov, Y. (1993). Multiple attractors, catastrophes and chaos in seasonally perturbed predator-prey communities. *Bulletin of Mathematical Biology* 55, 15–35.

Sarkar, R. R., Chattopadhayay, J. (2003). Occurence of planktonic blooms under environmental fluctuations and its possible control mechanism – mathematical models and experimental observations. *Journal of Theoretical Biology* 224, 501–516.

Satnoianu, R. A., Menzinger, M. (2000). Non-Turing stationary patterns in flow-distributed oscillators with general diffusion and flow rates. *Physical Review E* 62(1), 113–119.

Satnoianu, R. A., Menzinger, M., Maini, P. K. (2000). Turing instabilities in general systems. *Journal of Mathematical Biology* 41, 493–512.

Scheffer, M. (1991a). Fish and nutrients interplay determines algal biomass: a minimal model. *Oikos* 62, 271–282.

Scheffer, M. (1991b). Should we expect strange attractors behind plankton dynamics - and if so, should we bother? *Journal of Plankton Research* 13, 1291–1305.

Scheffer, M., Rinaldi, S., Kuznetsov, Y. A., van Nes, E. H. (1997). Seasonal dynamics of daphnia and algae explained as a periodically forced predator-prey system. *Oikos* 80, 519–532.

Sherratt, J. A., Lewis, M. A., Fowler, A. C. (1995). Ecological chaos in the wake of invasion. *Proceedings of the National Academy of Sciences of the United States of America* 92, 2524–2528.

Steele, J., Henderson, E. W. (1992). A simple model for plankton patchiness. *Journal of Plankton Research* 14, 1397–1403.

Steffen, E., Malchow, H., Medvinsky, A. B. (1997). Effects of seasonal perturbation on a model plankton community. *Environmental Modeling and Assessment* 2, 43–48.

Suttle, C. A., Chan, A. M. (1993). Marine cyanophages infecting oceanic and coastal strains of *Synechococcus*: abundance, morphology, cross-infectivity and growth characteristics. *Marine Ecology Progress Series* 92, 99–109.

Suttle, C. A., Chan, A. M., Cottrell, M. T. (1990). Infection of phytoplankton by viruses and reduction of primary productivity. *Nature* 347, 467–469.

Suttle, C. A., Chan, A. M., Feng, C., Garza, D. R. (1993). Cyanophages and sunlight: a paradox. In *Trends in microbial ecology* (Guerrero, R., Pedros-Alio, C., eds.). Spanish Society for Microbiology, Barcelona, pp. 303–307.

Tarutani, K., Nagasaki, K., Yamaguchi, M. (2000). Viral impacts on total abundance and clonal composition of the harmful bloom-forming phytoplankton *Heterosigma akashiwo*. *Applied and Environmental Microbiology* 66(11), 4916–4920.

Thomas, J. W. (1995). *Numerical partial differential equations: Finite difference methods*. Springer-Verlag, New York.

Truscott, J. E., Brindley, J. (1994). Ocean plankton populations as excitable media. *Bulletin of Mathematical Biology* 56, 981–998.

van Etten, J. L., Lane, L. C., Meints, R. H. (1991). Viruses and viruslike particles of eukaryotic algae. *Microbiological Reviews* 55, 586–620.

Wilcox, R., Fuhrman, J. (1994). Bacterial viruses in coastal seawater: lytic rather than lysogenic production. *Marine Ecology Progress Series* 114, 35–45.

Wilhelm, S. W., Suttle, C. A. (1999). Viruses and nutrient cycles in the sea. *BioScience* 49(10), 781–788.

Wommack, K. E., Colwell, R. R. (2000). Virioplankton: viruses in aquatic ecosystems. *Microbiology and Molecular Biology Reviews* 64(1), 69–114.

Zhdanov, V. P. (2003). Propagation of infection and the predator-prey interplay. *Journal of Theoretical Biology* 225, 489–492.

Chapter 6

Strange periodic attractors in a prey-predator system with infected prey

Frank M. Hilker and Horst Malchow

Institute of Environmental Systems Research, Department of Mathematics and Computer Science, University of Osnabrück, 49069 Osnabrück, Germany

November 30, 2004.

Abstract

A prey-predator model of phytoplankton-zooplankton dynamics is considered for the case of viral infection of the phytoplankton population. The phytoplankton population is split into a susceptible (S) and an infected (I) part. Both parts grow logistically, limited by a common carrying capacity. Zooplankton (Z) is grazing on susceptibles and infected, following a Holling-type II functional response. The analysis of the S-I-Z differential equations yields several semitrivial stationary states, among them two saddle-foci, and the sudden (dis-)appearance of a continuum of degenerated nontrivial equilibria. Along this continuum line, the equilibria undergo a fold-Hopf (zero-pair) bifurcation. The continuum only exists in the bifurcation point of the saddle-foci. Especially interesting is the emergence of strange periodic attractors, stabilizing themselves after a repeated torus-like oscillation.

6.1   Introduction

Conceptual prey-predator models have often and successfully been used to model phytoplankton-zooplankton interactions and to elucidate mechanisms of spatio-temporal pattern formation like patchiness and blooming (Segel and Jackson, 1972; Steele and Henderson, 1981; Scheffer, 1991; Pascual, 1993; Malchow, 1993).

Not so much known is about marine viruses and their role in aquatic ecosystems and the species that they infect, cf. Fuhrman (1999). Suttle et al. (1990) have experimentally shown that the viral disease can infect bacteria and phytoplankton in coastal water. There is some evidence that viral infection might accelerate the termination of phytoplankton blooms (Jacquet et al., 2002; Gastrich et al., 2004). However, despite the increasing number of reports, the role of viral infection in the phytoplankton population is still far from understood.

Mathematical models of the dynamics of virally infected phytoplankton populations are rare as well, the already classical publication is by Beltrami and Carroll (1994). More recent work is of Chattopadhyay and Pal (2002). The latter deal with lytic infections and mass action incidence functions. Malchow et al. (2004, 2005) observed oscillations and waves in a phytoplankton-zooplankton system with Holling-type II and III grazing under lysogenic viral infection and frequency-dependent transmission. The latter is also called proportionate mixing or standard incidence (Nold, 1980; Hethcote, 2000; McCallum et al., 2001).

In this paper, we investigate the local dynamics of phytoplankton with lytic infection and frequency-dependent transmission as well as zooplankton with Holling-type II grazing.

6.2   The mathematical model

The Scheffer model (Scheffer, 1991) for the prey-predator dynamics of phytoplankton P and zooplankton Z at time t reads in dimensionless quantities

$$\frac{\mathrm{d}P}{\mathrm{d}t} = rP(1-P) - \frac{aP}{1+bP}Z, \tag{6.1}$$

$$\frac{\mathrm{d}Z}{\mathrm{d}t} = \frac{aP}{1+bP}Z - m_3 Z. \tag{6.2}$$

There is logistic growth of the phytoplankton with intrinsic rate r and Holling-type II grazing with maximum rate a/b as well as natural mortality of zooplankton with rate m_3. The growth rate r is scaled as the ratio of local rate r_{loc} and a mean $\langle r \rangle$. The effects of nutrient supply and planktivorous fish are neglected because the focus of this paper is on the influence of the viral infection of phytoplankton. The phytoplankton population P is split into a susceptible part S and an infected portion I. Zooplankton is simply renamed to Z. It grazes on susceptible and infected phytoplankton. Then, the model system reads for symmetric inter- and intraspecific competition of susceptibles and infected

$$\frac{\mathrm{d}S}{\mathrm{d}t} = r_1 S(1 - S - I) - \frac{aS}{1+b(S+I)}Z - \lambda\frac{SI}{S+I}, \tag{6.3a}$$

$$\frac{\mathrm{d}I}{\mathrm{d}t} = r_2 I(1 - S - I) - \frac{aI}{1+b(S+I)}Z + \lambda\frac{SI}{S+I} - m_2 I, \tag{6.3b}$$

$$\frac{\mathrm{d}Z}{\mathrm{d}t} = \frac{a(S+I)}{1+b(S+I)}Z - m_3 Z. \tag{6.3c}$$

Frequency-dependent transmission rate λ as well as an additional disease-induced mortality of infected (virulence) with rate m_2 are assumed. The intrinsic growth rates of susceptibles and infected are r_1 and r_2, respectively. In the case of lysogenic infection, it holds $0 \leq r_2 \leq r_1$, whereas in the case of lytic infection $r_2 \leq 0 \leq r_1$. Then, the first term on the right-hand side of eq. (6.3b) describes the losses due to natural mortality and competition.

6.3 The stationary dynamics

Now, it is searched for stationary and oscillatory solutions of system (6.3a–6.3c). To do that, it is simplified through a convenient transformation, then describing the dynamics of the total phytoplankton population $P = S + I$ and the prevalence $i = I/P$. The vector of population densities is $\mathbf{X} = \{P, i, Z\}$. The model equations read

$$\frac{\mathrm{d}\,P}{\mathrm{d}\,t} = [r_1(1-i) + r_2\,i](1-P)P - \frac{aP}{1+bP}Z - m_2\,i\,P, \qquad (6.4a)$$

$$\frac{\mathrm{d}\,i}{\mathrm{d}\,t} = [(r_2 - r_1)(1-P) + (\lambda - m_2)]\,(1-i)\,i, \qquad (6.4b)$$

$$\frac{\mathrm{d}\,Z}{\mathrm{d}\,t} = \frac{aP}{1+bP}Z - m_3 Z. \qquad (6.4c)$$

System (6.4a–6.4c) possesses the following (semi-)trivial equilibria

$$\mathbf{E} = \{P^S, i^S, Z^S\} \quad \text{with} \quad \left.\frac{\mathrm{d}\,P}{\mathrm{d}\,t}\right|_{\mathbf{X}=E} = \left.\frac{\mathrm{d}\,i}{\mathrm{d}\,t}\right|_{\mathbf{X}=E} = \left.\frac{\mathrm{d}\,Z}{\mathrm{d}\,t}\right|_{\mathbf{X}=E} = 0 :$$

1) $\mathbf{E}_{00} = \{0, 0, 0\}$.

   The trivial state is always unstable.

2) $\mathbf{E}_{01} = \{0, i_{01}^S, 0\}$ with $i_{01}^S = 1$.

   This disease-induced extinction of the total prey population occurs for $r_2 < m_2 < \lambda + r_2 - r_1$.

3) $\mathbf{E}_1 = \{P_1^S, 0, 0\}$ with $P_1^S = S = 1$.

   Only the susceptible prey species survive at their capacity for $\lambda < m_2$ and $m_3 > a/(1+b)$.

4) $\mathbf{E}_2 = \{P_2^S, i_2^S, 0\}$ with $P_2^S > 0$, $i_2^S > 0$.

   a) $\mathbf{E}_{21} = \{P_{21}^S, i_{21}^S, 0\}$ with $P_{21}^S = I = 1 - m_2/r_2$, $i_{21}^S = 1$.
      Only the infected survive for $m_2 < r_2$, $m_2 < (r_2/r_1)\,\lambda$ and $aP_{21}^S/(1+bP_{21}^S) < m_3$.

   b) $\mathbf{E}_{22} = \{P_{22}^S, i_{22}^S, 0\}$ with

   $$P_{22}^S = 1 - \frac{\lambda - m_2}{r_1 - r_2}, \quad i_{22}^S = \frac{r_1}{\lambda}\frac{\lambda - m_2}{r_1 - r_2}.$$

   The stability ranges of this solution can easily be found by some computer algebra tool. However, the expressions are rather lengthy and omitted here.

5) $\mathbf{E}_3 = \{P_3^S, 0, Z_3^S\}$ with

$$P_3^S = \frac{m_3}{a - m_3 b}, \quad Z_3^S = \frac{r_1}{a}(1 + bP_3^S)(1 - P_3^S).$$

The infected go extinct for too high virulence or too low transmission rate. The remaining textbook example of the P-Z prey-predator model is well studied. The solution can be a stable node or focus as well as, after a Hopf bifurcation, an unstable focus bound by a stable limit cycle.

Nontrivial equilibria do only exist for a special parameter combination. In this case, there is a continuum of stationary states:

6) $\mathbf{E}_{Z_4^S(i)} = \{P_4^S, i, Z_4^S(P_4^S, i)\}$ with $P_4^S > 0$, $0 < i < 1$, $Z_4^S(P_4^S, i) > 0$.

From eqs. (6.4b) and (6.4c) one finds the expressions

$$P_{41}^S = 1 - \frac{\lambda - m_2}{r_1 - r_2} \quad \text{and} \tag{6.5a}$$

$$P_{42}^S = \frac{m_3}{a - m_3 b}, \tag{6.5b}$$

which define two parallel planes independent of i and Z in $(P$-i-$Z)$ phase space. These planes are orthogonal to the $(P$-$Z)$ and parallel to the $(i$-$Z)$ plane. Both must coincide, i.e., the system parameters have strictly to obey the relation

$$1 - \frac{\lambda - m_2}{r_1 - r_2} = \frac{m_3}{a - m_3 b} = P_4^S. \tag{6.6}$$

From eq. (6.4a), one obtains the plane

$$Z = Z(P, i) = \frac{1 + bP}{a} \{ r_1(1 - P) + [(r_2 - r_1)(1 - P) - m_2] i \} \tag{6.7}$$

Then, all points $\mathbf{E}_{Z_4^S(i)}$ lying on the straight intersection line of planes (6.6) and (6.7), i.e.,

$$Z_4^S(P_4^S, i) = Z(P_4^S, i) \text{ with } P_4^S \text{ as in (6.6)}, \tag{6.8}$$

independent of i for $0 < i < 1$ are stationary states. Obviously, this line of stationary states is a heteroclinic connection between the semitrivial equilibria $\mathbf{E}_2$ and $\mathbf{E}_3$.

For $m_3 < [a(b - 1)] / [b(b + 1)]$, the P-Z subsystem has an unstable focus bound by a stable limit cycle. In the stationary case (6.6), numerical analysis shows that all equilibria $\mathbf{E}_{Z_4^S(i)}$ on line (6.8), including $\mathbf{E}_2$ and $\mathbf{E}_3$, are degenerated, say their third eigenvalue is zero. The upper part of line (6.8) consists of degenerated unstable foci. A fold-Hopf (zero-pair) bifurcation point (Kuznetsov, 1995; Nicolis, 1995) separates them from the lower part closer to the $(P$-$i)$ plane with degenerated stable foci. A corresponding numerical simulation, starting on the unstable upper branch, is shown in Fig. 6.1.

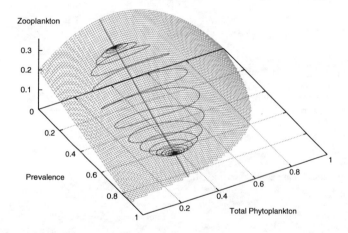

Figure 6.1: Stationary dynamics of system (6.4a-6.4c) with coexistence of all three populations. The trajectory starts at the upper unstable part of the line of stationary points, passes the fold-Hopf bifurcation point, and finally relaxes on the neutrally stable lower part. Parameters: $r_1 = 1$, $r_2 = 0$, $a = b = 5$, $\lambda = 4/5$ from eq. (6.6), $m_2 = 2/15$, $m_3 = 5/8$. Initial conditions: $P_0 = 0.335$, $i_0 = 1/5$, Z_0 from eq. (6.8). The straight line is the continuum of equilibria $\mathbf{E}_{Z_4^S(i)}$, lying on the shaded plane (6.7).

The closer the initial condition to the $(P\text{-}Z)$ plane the longer is the journey through phase space. The location of the final stationary state on the line strongly depends on the initial conditions, i.e., the final positions are only neutrally stable. This remains true in the case that the intersection point of (6.8) in the $P\text{-}Z$ subsystem is a stable focus or stable node, and the line (6.8) becomes a continuum of stable solutions. The latter case is not presented here.

The growth rate of infected r_2 has simply been set to zero. This choice describes cell lysis of infected phytoplankton cells and non-symmetric competition of infected and susceptibles, i.e., the infected still have an impact on the growth of susceptibles by shading and need for space, but not vice versa. Furthermore, m_2 stands now for an effective mortality (virulence plus natural mortality) of the infected.

6.4   Strange periodic attractors

The strong parameter relation (6.6) is surely not realistic. The probability to meet such a setting in a natural system is (almost) zero. Therefore, non-stationary situations will be simulated now, i.e., parameter settings when the planes (6.5a) and (6.5b) do not coincide and the intersection lines with plane (6.7) are not anymore stationary.

At first, the virulence is increased. Numerical bifurcation and stability analysis shows that in this case $\mathbf{E}_2$ is a saddle-focus with a stable two-dimensional manifold and an unstable one-dimensional manifold. In the $(P\text{-}Z)$ plane, $\mathbf{E}_3$ is also a saddle-focus, but with an unstable two-dimensional manifold and a stable one-dimensional manifold. In Fig. 6.2, the trajectory starts in the upper corner and approaches the lower end point of the right-hand line (6.5a) in the $(P\text{-}i)$ plane which is the semi-trivial stationary state $\mathbf{E}_2$. This is the mentioned saddle-focus with stable oscillation but unstable in direction of Z. Therefore, the trajectory is shot along the heteroclinic connection to the $(P\text{-}Z)$ plane and gets into the sphere of influence of the end point of the left-hand line (6.5b). This is, cf. above, also a semi-trivial saddle-focus, namely $\mathbf{E}_3$ with unstable oscillation and stable in Z direction. Hence, the trajectory bounces back, spirals down the lines and "tube-rides" up again and again, i.e., on the way up, it is "reinjected" (Nicolis, 1995) and tunnels through the two formed funnels. This is illustrated in Fig. 6.2a. It resembles the movement on a torus, where the centre hole of the torus is shrinked to a thin tube. However, the precessing trajectory gets "phase-locked" and finally, for long times, approaches a periodic attractor, see also Langford (1983, pp. 233). This is shown in Fig. 6.2b. The oscillation takes place in a plane which is orthogonal to the $(i\text{-}Z)$ plane. The attractor surrounds the two intersection points of the two lines of non-stationary points and the plane of oscillation.

For further illustration also the temporal development of the total prey density and its unfolded next-maximum map are given in Figs. 6.3. The latter resembles a damped oscillation.

The location of the periodic attractor is independent of the initial conditions like in the case of a limit cycle. For further increasing values of m_2, the behaviour of the system becomes simpler. The distinct funnel formation disappears and the periodic attractor is stabilizing faster and faster. For too high virulence,

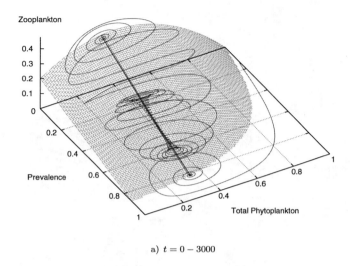

a) $t = 0 - 3000$

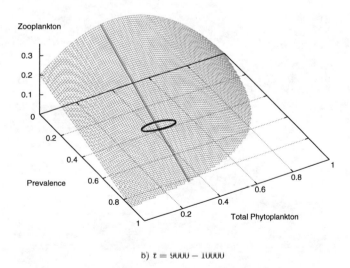

b) $t = 9000 - 10000$

Figure 6.2: Non-stationary dynamics of system (6.4a-6.4c) with coexistence of all three populations, cf. text. Parameters: $m_2 = 7/50$, all others like in Fig. 6.1. Inital condition: $P_0 = 1.01$, $i_0 = 0.1$, $Z_0 = 0.001$.

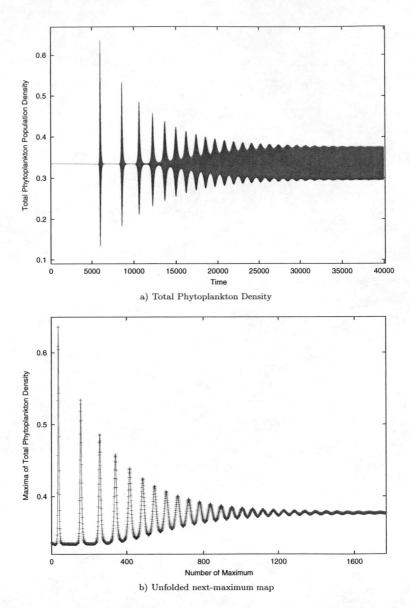

a) Total Phytoplankton Density

b) Unfolded next-maximum map

Figure 6.3: Non-stationary dynamics of system (6.4a-6.4c) with coexistence of all three populations, cf. text. Parameters: $m_2 = 27/200$, all others like in Fig. 6.1. Inital condition: $P_0 = 1/3$, $i_0 = 0.825$, $Z_0 = 0.0015$.

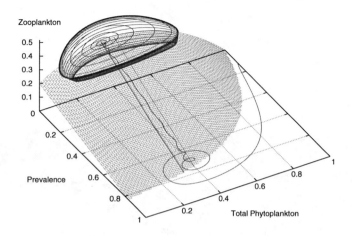

a) $m_2 = 1/5$, coexistence of all three populations

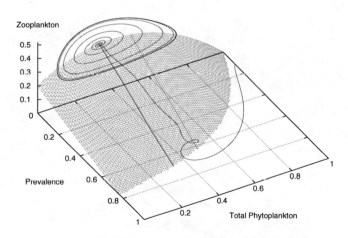

b) $m_2 = 3/10$, extinction of infected

Figure 6.4: Dynamics of system (6.4a-6.4c), cf. text. Parameters like in Fig. 6.1 except for m_2. Inital condition: $P_0 = 1.01$, $i_0 = 0.1$, $Z_0 = 0.001$.

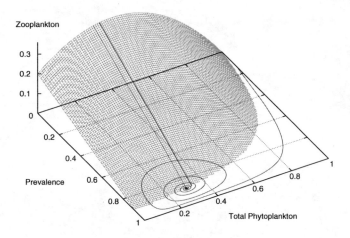

Figure 6.5: Dynamics of system (6.4a-6.4c), cf. text. Parameters: $m_2 = 0.1$, all others like in Fig. 6.1. Inital condition: $P_0 = 1.01$, $i_0 = 0.1$, $Z_0 = 0.001$.

the infected go extinct and the system oscillates in the P-Z subsystem. This is illustrated in Figs. 6.4.

For virulences below the stationary value given in Fig. 6.1, $\mathbf{E}_2$ becomes a stable and $\mathbf{E}_3$ an unstable focus, respectively. Numerical simulations yield that zooplankton dies out and the dynamics relaxes to $\mathbf{E}_2$ in the P-i subsystem, cf. Fig. 6.5.

In summary, this means that $\mathbf{E}_2$ has undergone a bifurcation from an unstable saddle-focus to a stable focus. At this bifurcation point, the continuum of degenerated nontrivial equilibria $\mathbf{E}_{Z_4^s(i)}$ simultaneously appeared. A zero-pair Hopf bifurcation took place along this continuum line.

6.5 Conclusion

A conceptual biomass-based model of phytoplankton-zooplankton prey-predator dynamics has been investigated for temporal pattern formation in a deterministic environment. It has been assumed that the phytoplankton is partly virally infected. Holling-type II zooplankton grazing has been considered and simulations have been performed for the case of lytic infection.

The dynamics of the system is surprisingly complex. Especially interesting is the coexistence of all populations on a periodic attractor that stabilizes itself under non-stationary parameter settings as well as the interplay of two saddle-foci that are connected by a heteroclinic orbit. Forthcoming work has to investigate the importance of this phenomenon for pattern formation scenarios in the spatially distributed system and under environmental noise.

References

Beltrami, E., Carroll, T. O. (1994). Modelling the role of viral disease in recurrent phytoplankton blooms. *Journal of Mathematical Biology* 32, 857–863.

Chattopadhyay, J., Pal, S. (2002). Viral infection on phytoplankton-zooplankton system – a mathematical model. *Ecological Modelling* 151, 15–28.

Fuhrman, J. A. (1999). Marine viruses and their biogeochemical and ecological effects. *Nature* 399, 541–548.

Gastrich, M. D., Leigh-Bell, J. A., Gobler, C. J., Anderson, O. R., Wilhelm, S. W., Bryan, M. (2004). Viruses as potential regulators of regional brown tide blooms caused by the alga, *Aureococcus anophagefferens*. *Estuaries* 27(1), 112–119.

Hethcote, H. W. (2000). The mathematics of infectious diseases. *SIAM Review* 42(4), 599–653.

Jacquet, S., Heldal, M., Iglesias Rodriguez, D., Larsen, A., Wilson, W., Bratbak, G. (2002). Flow cytometric analysis of an *Emiliana huxleyi* bloom terminated by viral infection. *Aquatic Microbial Ecology* 27, 111–124.

Kuznetsov, Y. A. (1995). *Elements of applied bifurcation theory*. Springer-Verlag, Berlin.

Langford, W. F. (1983). A review of interactions of Hopf and steady-state bifurcations. In *Nonlinear dynamics and turbulence* (Barenblatt, G. I., Iooss, G., Joseph, D. D., eds.). Interaction of Mechanics and Mathematics Series, Pitman, Boston, pp. 215–237.

Malchow, H. (1993). Spatio-temporal pattern formation in nonlinear nonequilibrium plankton dynamics. *Proceedings of the Royal Society of London B* 251, 103–109.

Malchow, H., Hilker, F. M., Petrovskii, S. V., Brauer, K. (2004). Oscillations and waves in a virally infected plankton system: Part I: The lysogenic stage. *Ecological Complexity* 1(3), 211–223.

Malchow, H., Hilker, F. M., Sarkar, R. R., Brauer, K. (2005). Spatiotemporal patterns in an excitable plankton system with lysogenic viral infection. *Mathematical and Computer Modelling* (in press).

McCallum, H., Barlow, N., Hone, J. (2001). How should pathogen transmission be modelled? *Trends in Ecology & Evolution* 16(6), 295–300.

Nicolis, G. (1995). *Introduction to nonlinear science*. Cambridge University Press, Cambridge.

Nold, A. (1980). Heterogeneity in disease-transmission modeling. *Mathematical Biosciences* 52, 227–2402.

Pascual, M. (1993). Diffusion-induced chaos in a spatial predator-prey system. *Proceedings of the Royal Society of London B* 251, 1–7.

Scheffer, M. (1991). Fish and nutrients interplay determines algal biomass: a minimal model. *Oikos* 62, 271–282.

Segel, L. A., Jackson, J. L. (1972). Dissipative structure: an explanation and an ecological example. *Journal of Theoretical Biology* 37, 545–559.

Steele, J., Henderson, E. W. (1981). A simple plankton model. *The American Naturalist* 117, 676–691.

Suttle, C. A., Chan, A. M., Cottrell, M. T. (1990). Infection of phytoplankton by viruses and reduction of primary productivity. *Nature* 347, 467–469.

Chapter 7

Oscillations and waves in a virally infected plankton system Part II: Transition from lysogeny to lysis

Horst Malchow[1], Frank M. Hilker[1], Michel Langlais[2] and Sergei V. Petrovskii[3]

[1] Institute for Environmental Systems Research, Department of Mathematics and Computer Science, University of Osnabrück, D-49069 Osnabrück, Germany

[2] UMR CNRS 5466, Mathématiques Appliquées de Bordeaux, case 26, Université Victor Segalen Bordeaux 2, 146 rue Léo Saignat, 33076 Bordeaux Cedex, France

[3] Shirshov Institute of Oceanology, Russian Academy of Sciences, Nakhimovsky Prospekt 36, Moscow 117218, Russia

May 31, 2005.

Abstract

A model of phytoplankton-zooplankton prey-predator dynamics is considered for the case of viral infection of the phytoplankton population. The phytoplankton population is split into a susceptible (S) and an infected (I) part. Both parts grow logistically, limited by a common carrying capacity. Zooplankton (Z) as a Holling-type II predator is grazing on susceptibles and infected. The local and spatial analyses of the S-I-Z model with lysogenic infection have been presented in a previous paper (Malchow et al., Ecol. Complexity 1(3), 211-223, 2004b). This lysogenic stage is rather sensitive to environmental variability. Therefore, the effect of a transition from lysogeny to lysis is investigated here. The replication rate of the infected species instantaneously falls to zero. A deterministic and a more realistic stochastic scenario are described. The spatiotemporal behaviour, modelled by deterministic and stochastic reaction-diffusion equations, is numerically determined. It is shown that the extinction risk of the infected is rather high in the deterministic system whereas the environmental noise enhances their chance of spatial spread and survival.

7.1   Introduction

There is not so much known about marine viruses and their role in aquatic ecosystems and the species that they infect, cf. Fuhrman (1999). Suttle et al. (1990) have experimentally shown that the viral disease can infect bacteria and phytoplankton in coastal water. There is some evidence that viral infection might accelerate the termination of phytoplankton blooms (Jacquet et al., 2002; Gastrich et al., 2004). However, despite the increasing number of reports, the role of viral infection in the phytoplankton population is still far from understood.

Mathematical models of the growth and interactions of virally infected phytoplankton populations are correspondingly rare. The already classical publication is by Beltrami and Carroll (1994). More recent work is of Chattopadhyay and Pal (2002). The latter deal with lytic infections and mass action incidence functions. Malchow et al. (2004b, 2005) observed oscillations and waves in a phytoplankton-zooplankton system with Holling-type II and III grazing under lysogenic viral infection and frequency-dependent transmission. The latter is also called proportionate mixing or standard incidence (Nold, 1980; Hethcote, 2000; McCallum et al., 2001). Hilker and Malchow (submitted) have provided a detailed mathematical and numerical analysis of the local model for lysogenic and lytic infections.

The understanding of the importance of lysogeny is just at the beginning (Wilcox and Fuhrman, 1994; Jiang and Paul, 1998; McDaniel et al., 2002; Ortmann et al., 2002). Contrary to lytic infections with destruction and without reproduction of the host cell, lysogenic infections are a strategy whereby viruses integrate their genome into the host's genome. As the host reproduces and duplicates its genome, the viral genome reproduces, too. The lysogenic replication cycle is rather sensitive to environmental fluctuations and, finally, switches to the lytic cycle with death of the host cell.

In this paper, we focus on modelling this switch for proportionate mixing incidence function (frequency-dependent transmission) and its consequences for the local and spatio-temporal dynamics of interacting phytoplankton and zooplankton. Furthermore, the impact of multiplicative noise (García-Ojalvo and Sancho, 1999; Allen, 2003; Anishenko et al., 2003) is investigated.

7.2   The basic mathematical model

As in Part I of this paper (Malchow et al., 2004b), the prey-predator dynamics of phytoplankton and zooplankton is described by a standard model (Steele and Henderson, 1981; Scheffer, 1991; Malchow, 1993; Pascual, 1993). The phytoplankton population is split into a susceptible part X_1 and an infected portion X_2. Their growth rates $r_{1,2}$ are scaled as the ratio of the local rates $r_{1,2}^{loc}$ and the spatial mean $\langle r \rangle$. Zooplankton X_3 is a Holling-type II grazer with maximum grazing rate a/b and natural mortality m_3. Then, the model system reads for symmetric inter- and intraspecific competition of susceptibles and infected in time t and two horizontal dimensions $\vec{r} = \{x, y\}$

$$\frac{\partial X_i(\vec{r}, t)}{\partial t} = f_i\left[\mathbf{X}(\vec{r}, t)\right] + d\,\Delta X_i(\vec{r}, t)\,, \ i = 1, 2, 3\,; \tag{7.1}$$

where

$$f_1 = r_1 X_1\left(1 - X_1 - X_2\right) - \frac{aX_1}{1 + b\left(X_1 + X_2\right)}X_3 - \lambda\frac{X_1 X_2}{X_1 + X_2}\,, \tag{7.1a}$$

$$f_2 = r_2 X_2\left(1 - X_1 - X_2\right) - \frac{aX_2}{1 + b\left(X_1 + X_2\right)}X_3 + \lambda\frac{X_1 X_2}{X_1 + X_2} - m_2 X_2\,, \tag{7.1b}$$

$$f_3 = \frac{a(X_1 + X_2)}{1 + b(X_1 + X_2)}X_3 - m_3 X_3\,. \tag{7.1c}$$

All quantities are dimensionless. The diffusion coefficient d describes eddy diffusion. Therefore, it must be equal for both species. Proportionate mixing with transmission coefficient λ as well as an additional disease-induced mortality of infected (virulence) with rate m_2 are assumed. The vector of population densities is $\mathbf{X} = \{X_1, X_2, X_3\}$. In the case of lytic infection, the first term on the right-hand side of eq. (7.1b) would describe the losses due to natural mortality and competition. At first, lysogenic infections with $r_1 \geq r_2 > 0$ will be considered. The growth rate of susceptibles is often higher than that of infected (Suttle et al., 1990). Secondly, the lysogenic viral replication cycle switches to the lytic one due to its high sensitivity to environmental fluctuations. In order to model the latter, multiplicative noise is introduced in eqs. (7.1), i.e.,

$$\frac{\partial X_i(\vec{r}, t)}{\partial t} = f_i\left[\mathbf{X}(\vec{r}, t)\right] + d\,\Delta X_i(\vec{r}, t) + \omega_i\left[\mathbf{X}(\vec{r}, t)\right]\cdot\xi_i(\vec{r}, t)\,, \ i = 1, 2, 3\,; \tag{7.2}$$

where $\xi_i(\vec{r}, t)$ is a spatiotemporal white Gaussian noise, i.e., a random Gaussian field with zero mean and delta correlation

$$\langle\xi_i(\vec{r}, t)\rangle = 0\,, \ \ \langle\xi_i(\vec{r}_1, t_1)\,\xi_i(\vec{r}_2, t_2)\rangle = \delta(\vec{r}_1 - \vec{r}_2)\,\delta(t_1 - t_2)\,, \ i = 1, 2, 3\,. \tag{7.2a}$$

$\omega_i\left[\mathbf{X}(\vec{r}, t)\right]$ is the density-dependent noise intensity. The stochastic modelling of population dynamics requires this density dependence, i.e., multiplicative noise. Throughout this paper, it is chosen

$$\omega_i\left[\mathbf{X}(\vec{r}, t)\right] = \omega X_i(\vec{r}, t)\,, \ i = 1, 2, 3\,; \ \omega = \text{const.} \tag{7.2b}$$

7.3 The local dynamics with deterministic switch

The local dynamics ($d = 0$) has been described by Malchow et al. (2004b) for the lysogenic case and $r_1 = r_2$. A general and more detailed analysis has been presented by Hilker and Malchow (submitted). Their results will not be listed here. The interest is rather in the spatiotemporal dynamics, starting from a system with lysogenic and then switching to lytic infection of the prey. Only one local example for such a switch is drawn in Fig. 7.1. After the transition, there is no further replication of infected. A more technical assumption for the simulation is that the remaining natural mortality of the infected is added

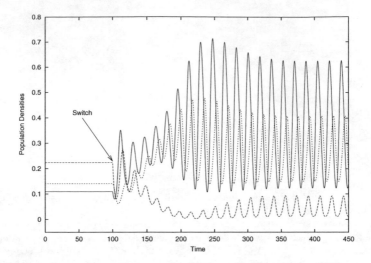

Figure 7.1: Non-oscillating endemic state with lysogenic infection switching at t=100 to an oscillating endemic state with lytic infection. The growth rate of infected r_2 switches from $r_2^{max} = 0.4$ to zero, the virulence m_2 from $m_2^{min} = 0.2$ to $m_2^{max} = 0.3$ and the transmission rate λ from $\lambda^{min} = 0.6$ to $\lambda^{max} = 0.9$, cp. text. Other parameter values: $r_1 = 1$, $a = b = 5$, $m_3 = 0.625$. Susceptibles are plotted with solid line, infected with dashed line and zooplankton with dots.

to the virulence, leading to an higher effective mortality of the infected, i.e., the parameter m_2 increases. Furthermore, the lytic cycle generates many more viruses, i.e., the transmission rate λ increases as well. And, finally, the intraspecific competition of the dying infected phytoplankton cells vanishes whereas the interspecific competition of susceptibles and infected becomes non-symmetric, i.e., the dead and dying infected still influence the growth of the susceptibles and contribute to the carrying capacity but not vice versa.

As to be expected, the switch from lysogeny to lysis results in a much lower mean abundance of infected though endemicity is still stable. However, the system responds rather sensitively to parameter changes, especially to variations of virulence and transmission rate and the infected can easily go extinct. As in the preceding paper, multiplicative noise supports the survival of the endangered species, i.e., there is always some probability to survive in a noisy environment while the deterministic setting inevitably leads to extinction. This is not further illustrated here.

7.4   The spatiotemporal dynamics

Now, we consider the spatiotemporal dynamics of the plankton model (7.2), i.e., zooplankton, grazing on susceptible and virally infected phytoplankton, under the influence of environmental noise and diffusing in horizontally two-dimensional space.

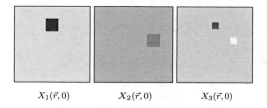

$$X_1(\vec{r}, 0) \qquad X_2(\vec{r}, 0) \qquad X_3(\vec{r}, 0)$$

Figure 7.2: Initial conditions for all spatiotemporal simulations.

7.4.1 Numerical methods, boundary and initial conditions

The diffusion terms have been integrated using the semi-implicit Peaceman-Rachford alternating direction scheme (Peaceman and Rachford Jr., 1955), cf. Thomas (1995). For the interactions and the Stratonovich integral of the noise terms, the explicit Euler-Maruyama scheme has been applied (Maruyama, 1955), cf. Kloeden and Platen (1999); Higham (2001). The time step of the numerical scheme is $\Delta t = 0.01$. The spatial grid is quadratic and has 99×99 grid points with spacing $\Delta x = \Delta y = 1$.

Periodic boundary conditions have been chosen for all simulations.

The initial conditions are as follows: The space is filled with the non-oscillating endemic state $(X_1^S, X_2^S, X_3^S) = (0.109, 0.224, 0.141)$, cf. Fig. 7.1. Furthermore, there are two localized patches in space. They can be seen in Fig. 7.2. The grey scale changes from high population densities in black colour to vanishing densities in white.

One patch is located in the upper middle of the model area with susceptibles $X_1 = 0.550$, four grid points ahead of zooplankton $X_3 = 0.450$ in each direction. The infected are at X_2^S. In the other patch at the right, the infected $X_2 = 0.333$ are ahead of zooplankton $X_3 = 0.036$ whereas the susceptibles are at X_1^S. These initial conditions are the same for deterministic and stochastic simulations.

The chosen system parameters generate oscillations in the center of the patches. The latter act as leading centers for target pattern waves that collide and break up to spirals. Increasing noise blurs this (naturally unrealistic) patterning, cf. also Malchow et al. (2004a,b).

7.4.2 Deterministic switching from lysogeny to lysis

At first, switching begins in the area with the highest initial density of infected, i.e., in the right-hand patch. The growth rate of infected r_2 vanishes whereas virulence and natural mortality of infected add up to a higher effective virulence m_2 and also the transmission rate λ increases as described in sec. 7.3 It is assumed that these parameter changes propagate through space like a Fisher wave (1937). If an auxiliary quantity r_3 with Fisher dynamics is introduced,

$$\frac{\partial r_3(\vec{r}, t)}{\partial t} = r_3(1 - r_3) + d\Delta r_3 , \tag{7.3}$$

$$\text{then} \quad r_2(\vec{r}, t) = r_2^{max}(1 - r_3) , \tag{7.3a}$$

$$m_2(\vec{r}, t) = m_2^{min} + (m_2^{max} - m_2^{min})r_3 , \tag{7.3b}$$

$$\lambda(\vec{r}, t) = \lambda^{min} + (\lambda^{max} - \lambda^{min})r_3 . \tag{7.3c}$$

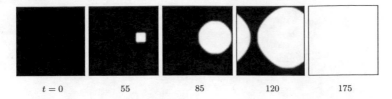

$t = 0$ 55   85   120   175

Figure 7.3: Spatial propagation of zero replication rate of the infected.

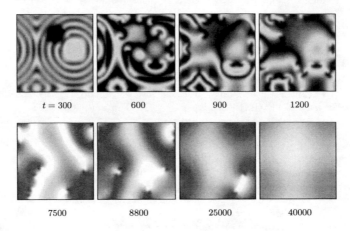

$t = 300$ 600   900   1200

7500   8800   25000   40000

Figure 7.4: Dynamics of susceptibles. No noise.

The initial conditions are $r_3 = 1$ in the right patch and zero elsewhere. For simplicity, the diffusivity d is assumed to be the same as for the populations. Its value of $d = 0.05$ has been chosen from Okubo's diffusion diagrams (1971) in order to model processes on a kilometer scale. The resulting spatial propagation of $r_2 = 0$ is drawn in white colour in Fig. 7.3.

The corresponding dynamics of susceptibles is plotted in Fig. 7.4. The presentation of susceptibles has been chosen because of richer contrast. The patterns of infected are complementary.

The break-up of concentric waves to a rather complex structure with spirals is nicely seen. In the long run, pinning-like behaviour of pairs of spirals is found. This effect is well-known from excitation waves in cardiac muscles, cf. the classical publications by Davidenko et al. (1992) and Pertsov et al. (1993). Here, the biological meaning remains unclear. The almost fixed pair-forming and rapidly rotating spirals approach each other extremely slowly, collide and burst. A weak multiplicative noise accelerates this process what is shown in Fig. 7.5. Stronger noise, i.e., higher environmental variability suppresses the generation of pins. Finally, the homogeneously oscillating endemic state remains.

The system parameters have been chosen to guarantee the survival of all three populations under deterministic conditions. The sample run for Fig. 7.5 with 5% noise also yields this final endemic coexistence, however, it should be noted that there is only a certain survival probability for all three, the lowest,

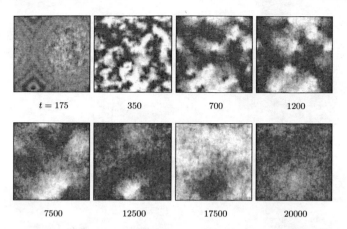

$t = 175$ 350   700   1200

7500   12500   17500   20000

Figure 7.5: Dynamics of susceptibles. Noise intensities $\omega_1=\omega_2=\omega_3=0.05$.

of course, for the infected after the switch to cell lysis.

7.4.3   Stochastic switching

The deterministic once-for-ever switching mechanism is highly unrealistic. One has to account for the certain fraction of the viruses which locally begins with the lysogenic replication cycle and then switches. This will be realized by a new auxiliary quantity r_3 with local bistable kinetics and multiplicative noise, i.e.

$$\frac{\mathrm{d}\,r_3(\vec{r},t)}{\mathrm{d}\,t} = (r_3 - r_3^{min})(r_3 - r_3^{crit})(r_3^{max} - r_3) + \omega_4 r_3 \cdot \xi(\vec{r},t) \ . \qquad (7.4)$$

The noise forces system (7.4) to switch between its stable stationary states r_3^{min} and r_3^{max} (Nitzan et al., 1974; Ebeling and Schimansky-Geier, 1980; Malchow and Schimansky-Geier, 1985). It is assumed that the replication rate of the infected switches accordingly, i.e.

$$\text{if } r_3 > r_3^{crit} \text{ then } m_2 = m_2^{min}, \lambda = \lambda^{min}, r_2 = r_2^{max} \text{ (lysogeny),} \qquad (7.4a)$$

$$\text{if } r_3 \le r_3^{crit} \text{ then } m_2 = m_2^{max}, \lambda = \lambda^{max}, r_2 = 0 \text{ (lysis).} \qquad (7.4b)$$

This as well as the temporal development of the spatial mean and the spatial-temporal pattern of r_2 are drawn in Figs. 7.6 and 7.7. Initially, the whole system is in the lysogenic state (7.4a).

At first, the simulation is run with noisy switches of r_2, λ and m_2 but deterministic population dynamics. In this unrealistic setting, the pins can still be seen, cf. Fig. 7.8.

If also the population dynamics is subject to noise, the result becomes more realistic. The pins are suppressed and the plankton forms a rather complex noise-induced patchy structure, cf. Fig. 7.9.

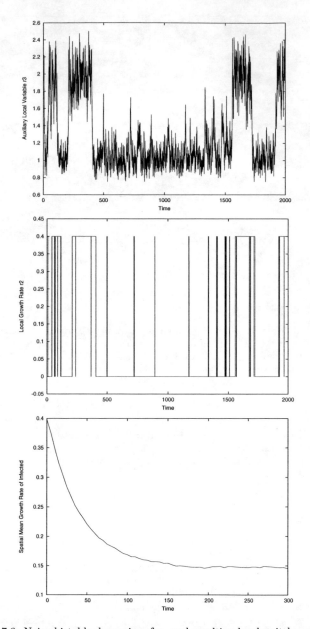

Figure 7.6: Noisy bistable dynamics of r_3 and resulting local switches of r_2 for $r_3^{max} = 2$, $r_3^{crit} = 1.5$, $r_3^{min} = 1$ and $\omega_4 = 0.1$. The spatial mean of r_2 decreases from the maximum as homogeneous initial condition to a value of approx. 0.15. The growth rate of infected r_2 switches from $r_2^{max} = 0.4$ to zero, the virulence m_2 from 0.2 to 0.3 and the transmission rate λ from 0.6 to 0.9, cp. text. Other parameter values: $r_1 = 1$, $a = b = 5$, $m_3 = 0.625$.

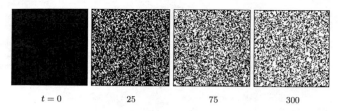

$t = 0$ 25   75   300

Figure 7.7: Spatio-temporal pattern of r_2.

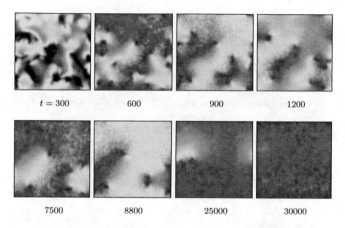

$t = 300$ 600   900   1200

7500   8800   25000   30000

Figure 7.8: Deterministic dynamics of susceptibles with noisy switch. $\omega_1 = \omega_2 = \omega_3 = 0$, $\omega_4 = 0.1$. The almost stationary spiral pairs do still exist.

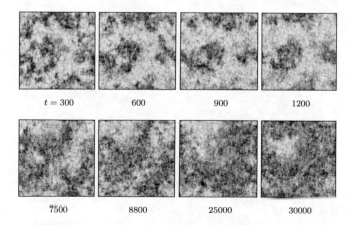

$t = 300$ 600   900   1200

7500   8800   25000   30000

Figure 7.9: Noisy dynamics of susceptibles and noisy switch. $\omega_i = 0.1$, $i = 1, 2, 3, 4$. The "pins" cannot be found anymore.

7.5   Conclusions

A conceptual biomass-based stochastic reaction-diffusion model of phytoplankton-zooplankton prey-predator dynamics has been investigated for temporal, spatial and spatio-temporal dissipative pattern formation in a deterministic and noisy environment, respectively. It has been assumed that the phytoplankton is partly virally infected and the virus switches from a lysogenic to a lytic replication cycle.

The logistic growth rate of lysogenically infected has been about 40% of the growth rate of susceptible phytoplankton. The local dynamics has been used to tune the system parameters in a way that the infected survive after switching from lysogeny to lysis.

In the spatial model, the switch has first been assumed to be deterministic, beginning at a certain position and then propagating like a Fisher wave. The populations have shown a complex wavy structure with formation of almost pinned pairs of spirals. The biological meaning of this pattern is unclear. The pair formation has disappeared for increasing noise.

Deterministic switching is an unrealistic once-for-ever mechanism. Therefore, a stochastic method has been developed and applied. Noisy switches and population dynamics generate a complex patchy spatio-temporal structure that is typical for natural plankton populations. The presented sample runs have led to a final endemic state with coexistence of susceptibles, infected and zooplankton like in the deterministic case. However, one should be aware that there are only certain (less than unity) survival probabilities for all three populations. There are good chances that the three-component systems switches to one of its subsystems. Noise has had not only an impact on the spatiotemporal coexistence of populations but it has been necessary to blur distinct artificial population structures like target patterns or spirals and to generate more realistic fuzzy patterns.

Acknowledgement

H.M. and F.M.H. acknowledge stimulating discussions with Veronika Huber on the switching mechanism from lysogeny to lysis in virally infected cells.

This work has been partially supported by Deutsche Forschungsgemeinschaft, grant no. 436 RUS 113/821.

References

Allen, L. J. S. (ed.) (2003). *An introduction to stochastic processes with applications to biology*. Pearson Education, Upper Saddle River NJ.

Anishenko, V. S., Astakov, V., Neiman, A. B., Vadivasova, T. E., Schimansky-Geier, L. (2003). *Nonlinear dynamics of chaotic and stochastic systems. Tutorial and modern developments*. Springer Series in Synergetics, Springer, Berlin.

Beltrami, E., Carroll, T. O. (1994). Modelling the role of viral disease in recurrent phytoplankton blooms. *Journal of Mathematical Biology* 32, 857–863.

Chattopadhyay, J., Pal, S. (2002). Viral infection on phytoplankton-zooplankton system – a mathematical model. *Ecological Modelling* 151, 15–28.

Davidenko, J. M., Pertsov, A. M., Salomonsz, R., Baxter, W., Jalife, J. (1992). Stationary and drifting spiral waves of excitation in isolated cardiac muscle. *Nature* 355, 349–351.

Ebeling, W., Schimansky-Geier, L. (1980). Nonequilibrium phase transitions and nucleation in reaction systems. In *Proceedings of the 6th International Conference on Thermodynamics*, Merseburg, pp. 95–100.

Fisher, R. A. (1937). The wave of advance of advantageous genes. *Annals of Eugenics* 7, 355–369.

Fuhrman, J. A. (1999). Marine viruses and their biogeochemical and ecological effects. *Nature* 399, 541–548.

García-Ojalvo, J., Sancho, J. M. (eds.) (1999). *Noise in spatially extended systems*. Institute for Nonlinear Science, Springer-Verlag, New York.

Gastrich, M. D., Leigh-Bell, J. A., Gobler, C. J., Anderson, O. R., Wilhelm, S. W., Bryan, M. (2004). Viruses as potential regulators of regional brown tide blooms caused by the alga, *Aureococcus anophageﬀerens*. *Estuaries* 27(1), 112–119.

Hethcote, H. W. (2000). The mathematics of infectious diseases. *SIAM Review* 42(4), 599–653.

Higham, D. (2001). An algorithmic introduction to numerical simulation of stochastic differential equations. *SIAM Review* 43(3), 525–546.

Hilker, F. M., Malchow, H. (submitted). Strange periodic attractors in a prey-predator system with infected prey.

Jacquet, S., Heldal, M., Iglesias-Rodriguez, D., Larsen, A., Wilson, W., Bratbak, G. (2002). Flow cytometric analysis of an *Emiliana huxleyi* bloom terminated by viral infection. *Aquatic Microbial Ecology* 27, 111–124.

Jiang, S. C., Paul, J. H. (1998). Significance of lysogeny in the marine environment: studies with isolates and a model of lysogenic phage production. *Microbial Ecology* 35(3), 235–243.

Kloeden, P. E., Platen, E. (1999). *Numerical solution of stochastic differential equations*. Springer, Berlin.

Malchow, H. (1993). Spatio-temporal pattern formation in nonlinear nonequilibrium plankton dynamics. *Proceedings of the Royal Society of London B* 251, 103–109.

Malchow, H., Hilker, F. M., Petrovskii, S. V. (2004a). Noise and productivity dependence of spatiotemporal pattern formation in a prey-predator system. *Discrete and Continuous Dynamical Systems B* 4(3), 705–711.

Malchow, H., Hilker, F. M., Petrovskii, S. V., Brauer, K. (2004b). Oscillations and waves in a virally infected plankton system: Part I: The lysogenic stage. *Ecological Complexity* 1(3), 211–223.

Malchow, H., Hilker, F. M., Sarkar, R. R., Brauer, K. (2005). Spatiotemporal patterns in an excitable plankton system with lysogenic viral infection. *Mathematical and Computer Modelling* (in press).

Malchow, H., Schimansky-Geier, L. (1985). *Noise and diffusion in bistable nonequilibrium systems*. No. 5 in Teubner-Texte zur Physik, Teubner-Verlag, Leipzig.

Maruyama, G. (1955). Continuous Markov processes and stochastic equations. *Rendiconti del Circolo Matematico di Palermo, Series II* 4, 48–90.

McCallum, H., Barlow, N., Hone, J. (2001). How should pathogen transmission be modelled? *Trends in Ecology & Evolution* 16(6), 295–300.

McDaniel, L., Houchin, L. A., Williamson, S. J., Paul, J. H. (2002). Lysogeny in *Synechococcus*. *Nature* 415, 496.

Nitzan, A., Ortoleva, P., Ross, J. (1974). Nucleation in systems with multiple stationary states. *Faraday Symposia of The Chemical Society* 9, 241–253.

Nold, A. (1980). Heterogeneity in disease-transmission modeling. *Mathematical Biosciences* 52, 227–2402.

Okubo, A. (1971). Oceanic diffusion diagrams. *Deep-Sea Research* 18, 789–802.

Ortmann, A. C., Lawrence, J. E., Suttle, C. A. (2002). Lysogeny and lytic viral production during a bloom of the cyanobacterium *Synechococcus* spp. *Microbial Ecology* 43, 225–231.

Pascual, M. (1993). Diffusion-induced chaos in a spatial predator-prey system. *Proceedings of the Royal Society of London B* 251, 1–7.

Peaceman, D. W., Rachford Jr., H. H. (1955). The numerical solution of parabolic and elliptic differential equations. *Journal of the Society for Industrial and Applied Mathematics* 3, 28–41.

Pertsov, A. M., Davidenko, J. M., Salomonsz, R., Baxter, W., Jalife, J. (1993). Spiral waves of excitation underlie reentrant activity in isolated cardiac muscle. *Circulation Research* 72, 631–650.

Scheffer, M. (1991). Fish and nutrients interplay determines algal biomass: a minimal model. *Oikos* 62, 271–282.

Steele, J., Henderson, E. W. (1981). A simple plankton model. *The American Naturalist* 117, 676–691.

Suttle, C. A., Chan, A. M., Cottrell, M. T. (1990). Infection of phytoplankton by viruses and reduction of primary productivity. *Nature* 347, 467–469.

Thomas, J. W. (1995). *Numerical partial differential equations: Finite difference methods*. Springer-Verlag, New York.

Wilcox, R., Fuhrman, J. (1994). Bacterial viruses in coastal seawater: lytic rather than lysogenic production. *Marine Ecology Progress Series* 114, 35–45.

Acknowledgements

First of all, I would like to express my gratitude to Horst Malchow. Since the beginning of my diploma studies in Osnabrück I have had the opportunity to learn from him, and during my PhD studies he has been much more than a supervisor. I have enjoyed collaborating with him in an extraordinarily stimulating, open and personal atmosphere – all the time being sure of his friendship and support.

I am also very thankful to Sergei Petrovskii, Michel Langlais, Hiromi Seno and Mark Lewis for our joint collaboration. In particular, I would like to thank Sergei Petrovskii for the discussions with him that I like and appreciate so much, Michel Langlais for his support and confidence in me, Hiromi Seno for his great and perfect hospitality during my stay in Japan and Mark Lewis for his stimulating suggestions.

There is a long list of persons to whom I am indebted for a couple of reasons: specific or general discussion, collaboration or support in various ways such as hosting me, invitations or arranging funding. To be precise again, I try to mention all of them (without any specific order): Nanako Shigesada and Fugo Takasu, Yoh Iwasa and Akira Sasaki, Norio Yamamura, Yuzo Hosono, Lutz Becks, Hartmut Arndt, Klaus Jürgens, Ezio Venturino, Frank Westerhoff, Klaus Brauer, Ram Rup Sarkar, Hans Joachim Poethke, Martin Hinsch, Thomas Hovestadt, Jean-Christophe Poggiale, Ulrike Feudel, Rafael Bravo de la Parra, Mimmo Iannelli, Wolfgang Alt, Andreas Deutsch, Lutz Schimansky-Geier and Bernd Blasius.

Furthermore, I acknowledge financial support from the following institutions: Japan Society for the Promotion of Science (JSPS), Institute for Advanced Study (IAS, Princeton NJ) / Park City Mathematics Institute (PCMI, Park City UT), Deutsche Forschungsgemeinschaft (DFG), Socrates/Erasmus Programm für Dozentenmobilität, Universitätsgesellschaft Osnabrück, Université Victor Segalen Bordeaux 2 and CIRM (Centre International des Rencontres Mathématique, Marseille).

Special thanks go to my colleagues at the University of Osnabrück for providing a highly pleasant atmosphere, including the students in my exercise courses for their interest and Andreas Focks and Anne Dietzel for their support as student assistants. Moreover, I express my gratitude to the persons in charge for the final report about my work.

Finally, I would like to thank all of my friends who had patience when I was busy with work or travelling again. I am especially grateful to my parents and to my sister Judith for all of their understanding and support.

Bibliography

Abramson, G., Kenkre, V. M., Yates, T. L., Parmenter, R. R. (2003). Traveling waves of infection in the hantavirus epidemics. *Bulletin of Mathematical Biology* 65, 519–534.

Ainseba, B. E., Fitzgibbon, W. E., Langlais, M., Morgan, J. J. (2002). An application of homogenization techniques to population dynamics models. *Communications on Pure and Applied Analysis* 1(1), 19–33.

Alexander, M. E., Moghadas, S. M. (2004). Periodicity in an epidemic model with a generalized non-linear incidence. *Mathematical Biosciences* 189, 75–96.

Allen, L. J. S. (ed.) (2003). *An introduction to stochastic processes with applications to biology*. Pearson Education, Upper Saddle River NJ.

Allen, L. J. S., Langlais, M., Phillips, C. J. (2003). The dynamics of two viral infections in a single host population with applications to hantavirus. *Mathematical Biosciences* 186, 191–217.

Andersen, M. C., Adams, H., Hope, B., Powell, M. (2004). Risk assessment for invasive species. *Risk Analysis* 24(4), 787–793.

Anderson, P. K., Cunningham, A. A., Patel, N. G., Morales, F. J., Epstein, P. R., Daszak, P. (2004). Emerging infectious diseases of plants: pathogen pollution, climate change and agrotechnology drivers. *Trends in Ecology & Evolution* 19(10), 535–544.

Anderson, R. M., Jackson, H. C., May, R. M., Smith, A. M. (1981). Population dynamics of foxes rabies in europe. *Nature* 289, 765–770.

Anderson, R. M., May, R. M. (1979). Population biology of infectious diseases: Part I. *Nature* 280, 361–367.

Anderson, R. M., May, R. M. (1991). *Infectious diseases of humans. Dynamics and control*. Oxford University Press, Oxford.

Andersson, H. (1998). Limit theorems for a random graph epidemic model. *Annals of Applied Probability* 8, 1331–1349.

Andow, D. A., Kareiva, P. M., Levin, S. A., Okubo, A. (1990). Spread of invading organisms. *Landscape Ecology* 4(177-188).

Andow, D. A., Kareiva, P. M., Levin, S. A., Okubo, A. (1993). Spread of invading species organisms: patterns of spread. In *Evolution of insect pest* (Kim, K. C., McPheron, B. A., eds.). Wiley, New York, pp. 219–242.

Andreasen, V., Lin, J., Levin, S. A. (1997). The dynamics of cocirculating influenza strains conferring partial cross-immunity. *Journal of Mathematical Biology* 35(825-842).

Anishenko, V. S., Astakov, V., Neiman, A. B., Vadivasova, T. E., Schimansky-Geier, L. (2003). *Nonlinear dynamics of chaotic and stochastic systems. Tutorial and modern developments*. Springer Series in Synergetics, Springer, Berlin.

Antonovics, J., Iwasa, Y., Hassell, M. P. (1995). A generalized model of parasitoid, venereal, and vector-based transmission processes. *The American Naturalist* 145, 661–675.

Arino, J., van den Driessche, P. (2003). A multi-city epidemic model. *Mathematical Population Studies* 10(3), 175–193.

Aronson, D. G., Weinberger, H. F. (1975). Nonlinear diffusion in population genetics, combustion, and nerve propagation. In *Partial differential equations and related topics* (Goldstein, J. A., ed.). No. 446 in Lecture Notes in Mathematics, Springer-Verlag, Berlin, pp. 5–49.

Aronson, D. G., Weinberger, H. F. (1978). Multidimensional nonlinear diffusion arising in population genetics. *Advances in Mathematics* 30(1), 33–76.

Arrigoni, F., Pugliese, A. (2002). Limits of a multi-patch SIS epidemic model. *Journal of Mathematical Biology* 45, 419–440.

Auger, P., Pontier, D. (1998). Fast game theory coupled to slow population dynamics: the case of domestic cat populations. *Mathematical Biosciences* 148, 65–82.

Bahi-Jaber, N., Langlais, M., Pontier, D. (2003). Behavioral plasticity and virus propagation: the FIV-cat population example. *Theoretical Population Biology* 64, 11–24.

Bailey, N. T. J. (1957). *The mathematical theory of epidemics*. Charles Griffin & Company Ltd, London.

Bailey, N. T. J. (1975). *The mathematical theory of infectious diseases and its applications*. Charles Griffin & Company Ltd, London, 2nd edition.

Barbour, A., Mollison, D. (1990). Epidemics and random graphs. In *Stochastic processes in epidemic theory* (Gabriel, J. P., Lefevre, C., Picard, P., eds.). Springer, New York, pp. 365–382.

Barthélemy, M., Barrat, A., Pastor-Satorras, R., Vespignani, A. (2005). Dynamical patterns of epidemic outbreaks in complex heterogeneous networks. *Journal of Theoretical Biology* 235, 275–288.

Becks, L., Hilker, F. M., Malchow, H., Jürgens, K., Arndt, H. (2005). Experimental demonstration of chaos in a microbial food web. *Nature* 435, 1226–1229.

Beltrami, E. (1989). A mathematical model of the brown tide. *Estuaries* 12, 13–17.

Beltrami, E. (1996). Unusual algal blooms as excitable systems: The case of "brown-tides". *Environmental Modeling and Assessment* 1, 19–24.

Beltrami, E., Carroll, T. O. (1994). Modelling the role of viral disease in recurrent phytoplankton blooms. *Journal of Mathematical Biology* 32, 857–863.

Bendinelli, M., Pistello, M., Lombardi, S., Poli, A., Gerzelli, C., Matteucii, D., Ceccherini-Nelli, L., Malvaldi, G., Tozzi, F. (1995). Feline Immunodeficiency Virus: an interesting model for AIDS studies and an important cat pathogen. *Clinical Microbiology Reviews* 8(1), 87–112.

Beretta, E., Kuang, Y. (1998). Modeling and analysis of a marine bacteriophage infection. *Mathematical Biosciences* 149, 57–76.

Bergh, O., Borsheim, K. Y., Bratbak, G., Heldal, M. (1989). High abundance of viruses found in aquatic environments. *Nature* 340, 467–468.

Boots, M., Sasaki, A. (2003). Parasite evolution and extinctions. *Ecology Letters* 6, 176–182.

Bowers, R. D., Turner, J. (1997). Community structure and the interplay between interspecific infection and competition. *Journal of Theoretical Biology* 187, 95–109.

Brandon, R. (1995). Retroviruses of cats: A review. *Australian Veterinary Practitioner* 25, 8–17.

Bratbak, G., Levasseur, M., Michand, S., Cantin, G., Fernandez, E., Heldal, M. (1995). Viral activity in relation to *Emiliania huxleyi* blooms: a mechanism of DMSP release? *Marine Ecology Progress Series* 128, 133–142.

Brauer, F. (1990). Models for the spread of universally fatal diseases. *Journal of Mathematical Biology* 28, 451–462.

Brauer, F., Castillo-Chavez, C. (2001). *Mathematical models in population biology and epidemiology.* Springer-Verlag, New York.

Britton, N. F. (1986). *Reaction-diffusion equations and their applications to biology.* Academic Press, London.

Brownlee, J. (1911). The mathematical theory of random migration and epidemic distribution. *Proceedings of the Royal Society of Edinburgh* 31, 262–289.

Brussard, C. P. D., Kempers, R. S., Kop, A. J., Riegman, R., Heldal, M. (1996). Virus-like particles in a summer bloom of *Emiliania huxleyi* in the North Sea. *Aquatic Microbial Ecology* 10(2), 105–113.

Busenberg, S., Cooke, K. (1993). *Vertically transmitted diseases. Models and dynamics.* No. 23 in Biomathematics, Springer-Verlag, Berlin.

Busenberg, S., van den Driessche, P. (1990). Analysis of a disease transmission model in a population with varying size. *Journal of Mathematical Biology* 28, 257–270.

Cantrell, R. S., Cosner, C. (2003). *Spatial ecology via reaction-diffusion equations.* Wiley, Chichester.

Capasso, V. (1993). *Mathematical structures of epidemics systems*. No. 97 in Lectures Notes in Biomathematics, Springer-Verlag, New York.

Capdevilla-Argüelles, L., Zilletti, B. (2005). *Issues in bioinvasion science*. Springer, Dordrecht.

Cappuccino, N. (2004). Allee effect in an invasive alien plant, pale swallow-wort *Vincetoxicum rossicum* (Asclepiadaceae). *Oikos* 106, 3–8.

Caraco, T., Glavanakov, S., Chen, G., Flaherty, J. E., Ohsumi, T. K., Szymanski, B. K. (2002). Stage-structured infection transmission and a spatial epidemic: a model for lyme disease. *The American Naturalist* 160(3), 348–359.

Case, T. J., Holt, R. D., McPeek, M. A., Keitt, T. H. (2005). The community context of species' borders: ecological and evolutionary perspectives. *Oikos* 108, 28–46.

Castillo-Chavez, C., Blower, S., van den Driessche, P., Kirschner, D., Yakubu, A.-A. (eds.) (2002a). *Mathematical approaches for emerging and re-emerging infectious diseases. An introduction*. Springer-Verlag, New York.

Castillo-Chavez, C., Blower, S., van den Driessche, P., Kirschner, D., Yakubu, A.-A. (eds.) (2002b). *Mathematical approaches for emerging and re-emerging infectious diseases. Models, methods and theory*. Springer-Verlag, New York.

Chattopadhyay, J., Arino, O. (1999). A predator-prey model with disease in the prey. *Nonlinear Analysis* 36, 747–766.

Chattopadhyay, J., Pal, S. (2002). Viral infection on phytoplankton-zooplankton system – a mathematical model. *Ecological Modelling* 151, 15–28.

Chattopadhyay, J., Pal, S., El Abdllaoui, A. (2003a). Classical predator-prey system with infection of prey population – a mathematical model. *Mathematical Methods in the Applied Sciences* 26, 1211–1222.

Chattopadhyay, J., Sarkar, R. R., Ghosal, G. (2002). Removal of infected prey prevent limit cycle oscillations in an infected prey-predator system – a mathematical study. *Ecological Modelling* 156, 113–121.

Chattopadhyay, J., Sarkar, R. R., Pal, S. (2003b). Dynamics of nutrient-phytoplankton interaction in the presence of viral infection. *BioSystems* 68, 5–17.

Choo, K., Williams, P. D., Day, T. (2003). Host mortality, predation and the evolution of virulence. *Ecology Letters* 6, 310–315.

Clark, J. S., Carpenter, S. R., Barber, M., Collins, S., Dobson, A., Foley, J. A., Lodge, D. M., Pascual, M., Pielke Jr., R., Pizer, W., Pringle, C., Reid, W. V., Rose, K. A., Sala, O., Schlesinger, W. H., Wall, D. H., Wear, D. (2001a). Ecological forecasts: an emerging imperative. *Science* 293, 657–660.

Clark, J. S., Lewis, M., Horvath, L. (2001b). Invasion by extremes: population spread with variation in dispersal and reproduction. *The American Naturalist* 157(5), 537–554.

Clavero, M., García-Berthou, E. (2005). Invasive species are a leading cause of animal extinctions. *Trends in Ecology & Evolution* 20(3), 110.

Clay, K. (2003). Parasites lost. *Nature* 421, 585–586.

Collier, T., Van Steenwyk, R. (2004). A critical evaluation of augmentative biological control. *Biological Control* 31, 245–256.

Costantino, R. F., Cushing, J. M., Dennis, B., Desharnais, R. A. (1995). Experimentally induced transitions in the dynamic behaviour of insect populations. *Nature* 375, 227–230.

Costantino, R. F., Desharnais, R. A., Cushing, J. M., Dennis, B. (1997). Chaotic dynamics in an insect population. *Science* 275, 389–391.

Courchamp, F., Chapuis, J.-L., Pascal, M. (2003). Mammal invaders on islands: Impact, control and control impact. *Biological Reviews* 78(3), 347–383.

Courchamp, F., Clutton-Brock, T., Grenfell, B. (1999a). Inverse density dependence and the Allee effect. *Trends in Ecology & Evolution* 14(10), 405–410.

Courchamp, F., Cornell, S. J. (2000). Virus-vectored immunocontraception to control feral cats on islands: a mathematical model. *Journal of Applied Ecology* 37, 903–913.

Courchamp, F., Langlais, M., Sugihara, G. (1999b). Cats protecting birds: modelling the mesopredator release effect. *Journal of Animal Ecology* 68, 282–292.

Courchamp, F., Langlais, M., Sugihara, G. (1999c). Control of rabbits to protect island birds from cat predation. *Biological Conservation* 89, 219–225.

Courchamp, F., Langlais, M., Sugihara, G. (2000a). Rabbits killing birds: modelling the hyperpredation process. *Journal of Animal Ecology* 69, 154–164.

Courchamp, F., Pontier, D., Langlais, M., Artois, M. (1995a). Population dynamics of Feline Immunodeficiency Virus within cat populations. *Journal of Theoretical Biology* 175(4), 553–560.

Courchamp, F., Pontier, P. (1994). Feline Immunodeficiency Virus: an epidemiological review. *Comptes Rendus de l'Académie des Sciences de Paris, Séries III, Sciences de la Vie* 317, 1123–1134.

Courchamp, F., Pontier, P., Fromont, E., Artois, M. (1995b). Impact of two feline retroviruses on natural populations of domestic cat. *Mammalia* 59(4), 589–598.

Courchamp, F., Say, L., Pontier, D. (2000b). Transmission of Feline Immunodeficiency Virus in a population of cats (*Felis catus*). *Wildlife Research* 27, 603–611.

Courchamp, F., Sugihara, G. (1999). Modeling the biological control of an alien predator to protect island species from extinction. *Ecological Applications* 9(1), 112–123.

Courchamp, F., Suppo, C., Fromont, E., Bouloux, C. (1995c). Dynamics of two feline retroviruses (FIV and FeLV) within populations of cats. *Proceedings of the Royal Society of London B* 264, 785–794.

Courchamp, F., Yoccoz, N. G., Artois, M., Pontier, D. (1998). At-risk individuals in Feline Immunodeficiency Virus epidemiology: Evidence from a multivariate approach in a natural population of domestic cats (*Felis catus*). *Epidemiology and Infection* 121, 227–236.

Crawley, M. J. (1987). What makes a community invasible? In *Colonization, succession and stability* (Gray, A. J., Crawley, M. J., Edwards, P. J., eds.). Blackwell Scientific, Oxford, pp. 429–453.

Cruickshank, I., Gurney, W. S. C., Veitch, A. R. (1999). The characteristics of epidemics and invasions with thresholds. *Theoretical Population Biology* 56, 279–292.

Cushing, J. M., Costantino, R. F., Dennis, B., Desharnais, R. A., Henson, S. M. (2003). *Chaos in ecology. Experimental nonlinear dynamics*. Academic Press, San Diego.

Davidenko, J. M., Pertsov, A. M., Salomonsz, R., Baxter, W., Jalífe, J. (1992). Stationary and drifting spiral waves of excitation in isolated cardiac muscle. *Nature* 355, 349–351.

Davis, H. G., Taylor, C. M., Civille, J. C., Strong, D. R. (2004). An Allee effect at the front of a plant invasion: *Spartina* in a Pacific estuary. *Journal of Animal Ecology* 92, 321–327.

Dawes, J. H. P., Gog, J. R. (2002). The onset of oscillatory dynamics in models of multiple disease strains. *Journal of Mathematical Biology* 45, 471–510.

de Castro, F., Bolker, B. (2005). Mechanisms of disease-induced extinction. *Ecology Letters* 8, 117–126.

de Jong, M. C. M., Diekman, O., Heesterbeek, H. (1995). How does transmission rate of infection depend on populations size? In *Epidemic models: Their structure and relation to data* (Mollison, D., ed.). The Newton Institute, Cambridge, UK, pp. 84–94.

Dennis, B. (1989). Allee effects: population growth, critical density, and the chance of extinction. *Natural Resource Modeling* 3, 481–538.

Dennis, B., Desharnais, R. A., Cushing, J. M., Costantino, R. F. (1995). Nonlinear demographic dynamics: mathematical models, statistical methods, and biological experiments. *Ecological Monographs* 65, 261–281.

Derrick, W. R., van den Driessche, P. (1993). A disease transmission model in a nonconstant population. *Journal of Mathematical Biology* 31, 495–512.

Derrick, W. R., van den Driessche, P. (2003). Homoclinic orbits in a disease transmission model with nonlinear incidence and nonconstant population. *Discrete and Continuous Dynamical Systems B* 3(2), 299–309.

Diekmann, O., de Jong, M. C. M., Metz, J. A. J. (1990). On the definition and the computation of the basic reproductive ratio R_0 in models for infectious diseases in heterogeneous populations. *Journal of Mathematical Biology* 28, 365–384.

Diekmann, O., Heesterbeek, J. A. P. (2000). *Mathematical epidemiology of infectious diseases. Model building, analysis and interpretation.* John Wiley & Son, New York.

Diekmann, O., Kretzschmar, M. (1991). Patterns in the effects of infectious diseases on population growth. *Journal of Mathematical Biology* 29, 539–570.

Diekmann, O., Montijn, R. (1982). Prelude to Hopf bifurcation in an epidemic model: analysis of the characteristic equation associated with a nonlinear Volterra integral equation. *Journal of Mathematical Biology* 14, 117–127.

Dietz, K., Schenzle, D. (1985). Proportionate mixing models for age-dependent infection transmission. *Journal of Mathematical Biology* 22, 117–120.

Drake, J. A., Mooney, H. A., di Castro, F., Groves, R. H., Kruger, F. J., Rejmánek, M., Williamson, M. (eds.) (1989). *Biological invasions: A global perspective.* No. 27 in SCOPE, Wiley, Chichester.

Drake, J. M. (2004). Allee effects and the risk of biological invasion. *Risk analysis* 24(4), 795–802.

Dunbar, S. R. (1983). Travelling wave solutions of diffusive Lotka-Volterra equations. *Journal of Mathematical Biology* 17, 11–32.

Dunbar, S. R. (1984). Travelling wave solutions of diffusive Lotka-Volterra equations: a heteroclinic connection in R^4. *Transactions of the American Mathematical Society* 286, 557–594.

Dunbar, S. R. (1986). Travelling waves in diffusive predator-prey equations: periodic orbits and point-to-periodic heteroclinic orbits. *SIAM Journal on Applied Mathematics* 46, 1057–1078.

Dwyer, G. (1994). Density dependence and spatial structure in the dynamics of insect pathogens. *The American Naturalist* 143(4), 533–562.

Ebeling, W., Schimansky-Geier, L. (1980). Nonequilibrium phase transitions and nucleation in reaction systems. In *Proceedings of the 6th International Conference on Thermodynamics*, Merseburg, pp. 95–100.

Edelstein-Keshet, L. (1988). *Mathematical models in biology.* Birkhäuser Mathematics Series, McGraw-Hill, New York.

Ehrlich, P. R. (1986). Which animal will invade? In *Ecology of biological invasions of North America and Hawaii* (Mooney, H. A., Drake, J. A., eds.). Springer-Verlag, New York, pp. 70–95.

Elton, C. S. (1958). *The ecology of invasions by animals and plants.* Methuen, London.

Enderle, J. D. (1980). *A stochastic communicable disease model with age-specific states and applications to measles.* Ph.D. thesis, Rensselaer Polytechnic Institute.

Ewald, P. W. (1994). *Ecology of infectious diseases.* Oxford University Press, Oxford.

Fagan, W. F., Bishop, J. G. (2000). Trophic interactions during primary succession: herbivores slow a plant reinvasion at Mount St. Helens. *The American Naturalist* 155, 238–251.

Fagan, W. F., Lewis, M. A., Neubert, M. G., van den Driessche, P. (2002). Invasion theory and biological control. *Ecology Letters* 5, 148–158.

Fife, P. C. (1979). *Mathematical aspects of reacting and diffusing systems.* No. 28 in Lecture Notes in Biomathematics, Springer-Verlag, Berlin.

Fife, P. C., McLeod, J. B. (1977). The approach of solutions of nonlinear diffusion equations to travelling wave solution. *Archive for Rational Mechanics and Analysis* 65, 335–361.

Fisher, R. A. (1937). The wave of advance of advantageous genes. *Annals of Eugenics* 7, 355–369.

Fitzgibbon, W. E., Langlais, M. (2003). A diffusive S.I.S. model describing the propagation of FIV. *Communication in Applied Analysis* 3, 387–404.

Fitzgibbon, W. E., Langlais, M., Morgan, J. J. (2001). A mathematical model of the spread of Feline Leukemia Virus (FeLV) through a highly heterogeneous spatial domain. *SIAM Journal on Mathematical Analysis* 33(3), 570–578.

Frantzen, J. (2000). Disease epidemics and plant competition: control of *Senecio vulgaris* with *Puccinia lagenophorae*. *Basic and Applied Ecology* 1(2), 141–148.

Freedman, H. I., Moson, P. (1990). Persistence definitions and their connections. *Proc. Amer. Math. Soc.* 109, 1025–1033.

Fromont, E., Pontier, D., Langlais, M. (1998). Dynamics of a feline retrovirus (FeLV) in host populations with variable spatial structure. *Proceedings of the Royal Society of London B* 265, 1097–1104.

Fromont, E., Pontier, D., Langlais, M. (2003). Disease propagation in connected host populations with density-dependent dynamics: the case of the Feline Leukemia Virus. *Journal of Theoretical Biology* 223, 465–475.

Fuhrman, J. A. (1999). Marine viruses and their biogeochemical and ecological effects. *Nature* 399, 541–548.

Gao, L. Q., Hethcote, H. W. (1992). Disease transmission models with density-dependent demographics. *Journal of Mathematical Biology* 30, 717–731.

Gao, L. Q., Mena-Lorca, J., Hethcote, H. W. (1995). Four SEI endemic models with periodicity and separatrices. *Mathematical Biosciences* 128, 157–184.

García-Ojalvo, J., Sancho, J. M. (eds.) (1999). *Noise in spatially extended systems*. Institute for Nonlinear Science, Springer-Verlag, New York.

Garrett, K. A., Bowden, R. L. (2002). An Allee effect reduces the invasive potential of *Tilletia indica*. *Phytopathology* 92(11), 1152–1159.

Gastrich, M. D., Leigh-Bell, J. A., Gobler, C. J., Anderson, O. R., Wilhelm, S. W., Bryan, M. (2004). Viruses as potential regulators of regional brown tide blooms caused by the alga, *Aureococcus anophagefferens*. *Estuaries* 27(1), 112–119.

Getz, W. M., Pickering, J. (1983). Epidemic models - thresholds and population regulation. *The American Naturalist* 121, 892–898.

Ghosh, M., Chandra, P., Sinha, P., Shukla, J. B. (2005). Modelling the spread of bacterial disease: effect of service providers from an environmentally degraded region. *Applied Mathematics and Computation* 160, 615–647.

Greenhalgh, D. (1990). An epidemic model with a density-dependent death rate. *IMA Journal of Mathematics Applied in Medicine & Biology* 7, 1–26.

Greenhalgh, D. (1992). Some results for an SEIR epidemic model with density dependence in the death rate. *IMA Journal of Mathematics Applied in Medicine & Biology* 9, 67–106.

Greenhalgh, D. (1997). Hopf bifurcation in epidemic models with a latent period and non-permanent immunity. *Mathematical and Computer Modelling* 25, 85–107.

Greenhalgh, D., Das, R. (1995). Modelling epidemics with variable contact rates. *Theoretical Population Biology* 47, 129–179.

Greenwood, M., Hill, A. T. B., Wilson, W. W. C. (1936). *Experimental epidemiology*. HMSO, London.

Grenfell, B. T., Bjørnstad, O. N., Kappey, J. (2001). Travelling waves and spatial hierarchies in measles epidemics. *Nature* 414, 716–723.

Grevstad, F. S. (1999). Factors influencing the change of population establishment: implications of release strategies in biocontrol. *Ecological Applications* 9(4), 1439–1447.

Gripenberg, G. (1980). Periodic solutions of an epidemic model. *Journal of Mathematical Biology* 10, 271–280.

Gurney, W. S. C., Nisbet, R. M. (1998). *Ecological dynamics*. Oxford University Press, New York.

Gurney, W. S. C., Veitch, A. R., Cruickshank, I., McGeachin, G. (1998). Circles and spirals: population persistence in a spatially explicit predator-prey model. *Ecology* 79(7), 2516–2530.

Hadeler, K. P., Freedman, H. I. (1989). Predator-prey populations with parasitic infection. *Journal of Mathematical Biology* 27, 609–631.

Hadeler, K. P., Rothe, F. (1975). Travelling fronts in nonlinear diffusion equations. *Journal of Mathematical Biology* 2, 251–263.

Hajek, A. E. (2004). *Natural enemies: An introduction to biological control.* Cambridge University Press, New York.

Hamer, W. M. (1906). Epidemic diseases in England. *Lancet* 1, 733–739.

Hanski, I., Turchin, P., Korpimäki, E., Henttonen, H. (1993). Population oscillations of boreal rodents: regulation by mustelid predators leads to chaos. *Nature* 364, 232–235.

Haraguchi, Y., Sasaki, A. (2000). The evolution of parasite virulence and transmission rate in a spatially structured population. *Journal of Theoretical Biology* 203, 85–96.

Hart, D. R., Gardner, R. H. (1997). A spatial model for the spread of invading organisms subject to competition. *Journal of Mathematical Biology* 35(8), 935–948.

Hastings, A. (2004). Transients: the key to long-term ecological understanding? *Trends in Ecology & Evolution* 19, 39–45.

Hastings, A., Cuddington, K., Davies, K. F., Dugaw, C. J., Elmendorf, S., Freestone, A., Harrison, S., Holland, M., Lambrinos, J., Malvadkar, U., Melbourne, B. A., Moore, K., Taylor, C., Thomson, D. (2005). The spatial spread of invasions: new developments in theory and evidence. *Ecology Letters* 8, 91–101.

Heger, T., Trepl, L. (2003). Predicting biological invasions. *Biological Invasions* 5, 313–321.

Hengeveld, R. (1989). *Dynamics of biological invasions.* Chapman and Hall, London.

Hethcote, H. W. (1973). Asymptotic behavior in a deterministic epidemic model. *Bulletin of Mathematical Biology* 35, 607–614.

Hethcote, H. W. (1976). Quantitative analyses of communicable disease models. *Mathematical Biosciences* 28, 335–356.

Hethcote, H. W. (1989). Three basic epidemiological models. In *Applied mathematical ecology* (Levin, S. A., Hallam, T. G., Gross, L. J., eds.). No. 18 in Biomathematics, Springer-Verlag, New York, pp. 119–144.

Hethcote, H. W. (1994). A thousand and one epidemic models. In *Frontiers in theoretical biology* (Levin, S. A., ed.). No. 100 in Lecture Notes in Biomathematics, Springer-Verlag, Berlin, pp. 504–515.

Hethcote, H. W. (2000). The mathematics of infectious diseases. *SIAM Review* 42(4), 599–653.

Hethcote, H. W., Lewis, M. A., van den Driessche, P. (1989). An epidemiological model with a delay and a non-linear incidence rate. *Journal of Mathematical Biology* 27, 49–64.

Hethcote, H. W., Stech, H. W., van den Driessche, P. (1981a). Nonlinear oscillations in epidemic models. *SIAM Journal on Applied Mathematics* 40(1), 1–9.

Hethcote, H. W., Stech, H. W., van den Driessche, P. (1981b). Periodicity and stability in epidemic models: A survey. In *Differential equations and applications in ecology, epidemics, and population problems* (Busenberg, S. N., Cooke, K. L., eds.). Academic Press, New York, pp. 65–82.

Hethcote, H. W., van den Driessche, P. (1991). Some epidemiological models with non-linear incidence. *Journal of Mathematical Biology* 29, 271.

Hethcote, H. W., Wang, W., Han, L., Ma, Z. (2004). A predator-prey model with infected prey. *Journal of Theoretical Biology* 66, 259–268.

Higham, D. (2001). An algorithmic introduction to numerical simulation of stochastic differential equations. *SIAM Review* 43(3), 525–546.

Hilker, F. M., Langlais, M., Petrovskii, S. V., Malchow, H. (accepted). A diffusive SI model with Allee effect and application to FIV. *Mathematical Biosciences* .

Hilker, F. M., Lewis, M. A., Seno, H., Langlais, M., Malchow, H. (2005). Pathogens can slow down or reverse invasion fronts of their hosts. *Biological Invasions* 7(5), 817–832.

Hilker, F. M., Malchow, H. (submitted). Strange periodic attractors in a prey-predator system with infected prey.

Holmes, E. E. (1997). Basic epidemiological concepts in a spatial context. In *Spatial ecology: The role of space in population dynamics and interspecific interactions* (Tilman, D., Kareiva, P., eds.). Princeton University Press, Princeton, pp. 111–136.

Holmes, E. E., Lewis, M. A., Banks, J. E., Veit, R. R. (1994). Partial differential equations in ecology: Spatial interactions and population dynamics. *Ecology* 75, 17–29.

Holt, R. D., Keitt, T. H., Lewis, M. A., Maurer, B. A., Taper, M. L. (2005). Theoretical models of species' borders: single species approaches. *Oikos* 108, 18–27.

Ilosono, Y. (1989). Singular perturbation analysis of travelling waves for diffusive Lotka-Volterra competitive models. In *Numerical and applied mathematics* (Brezinski, C., ed.). Baltzer, Basel, pp. 687–692.

Huang, J., Lu, G., Ruan, S. (2003). Existence of traveling wave solutions in a diffusive predator prey model. *Journal of Mathematical Biology* 46(2), 132–152.

Jacquet, S., Heldal, M., Iglesias-Rodriguez, D., Larsen, A., Wilson, W., Bratbak, G. (2002). Flow cytometric analysis of an *Emiliana huxleyi* bloom terminated by viral infection. *Aquatic Microbial Ecology* 27, 111–124.

Jeschke, J. M., Strayer, D. L. (2005). Invasion success of vertebrates in Europe and North America. *Proceedings of the National Academy of Sciences of the United States of America* 102, 7198–7202.

Jiang, S. C., Paul, J. H. (1998). Significance of lysogeny in the marine environment: studies with isolates and a model of lysogenic phage production. *Microbial Ecology* 35(3), 235–243.

Johansen, A. (1996). A simple model of recurrent epidemics. *Journal of Theoretical Biology* 178, 45–51.

Keane, R. M., Crawley, M. J. (2002). Exotic plant invasions and the enemy release hypothesis. *Trends in Ecology & Evolution* 17(4), 164–170.

Keitt, T. H., Lewis, M. A., Holt, R. D. (2001). Allee effects, invasion pinning, and species' borders. *The American Naturalist* 157, 203–216.

Kendall, D. G. (1965). Mathematical models of the spread of infection. In *Mathematics and computer science in biology and medicine.* HMSO, London, pp. 213–225.

Kermack, W. O., McKendrick, A. G. (1927). Contribution to the mathematical theory of epidemics, part I. *Proceedings of the Royal Society A* 115, 700–721.

Kermack, W. O., McKendrick, A. G. (1932). Contribution to the mathematical theory of epidemics. II - The problem of endemicity. *Proceedings of the Royal Society A* 138, 55–83.

Kermack, W. O., McKendrick, A. G. (1933). Contribution to the mathematical theory of epidemics. III - Further studies of the problem of endemicity. *Proceedings of the Royal Society A* 141, 94–122.

Kierstead, H., Slobodkin, L. B. (1953). The size of water masses containing plankton blooms. *Journal of Marine Research* XII, 141–147.

Kleczkowski, A., Grenfell, B. T. (1999). Mean-field-type equations for spread of epidemics: the 'small-world' model. *Physica A* 274, 355–360.

Kloeden, P. E., Platen, E. (1992). *Numerical solution of stochastic differential equations.* Springer, Berlin.

Kloeden, P. E., Platen, E. (1999). *Numerical solution of stochastic differential equations.* Springer, Berlin.

Kloeden, P. E., Platen, E., Schurz, H. (2002). *Numerical solution of SDE through computer experiments.* Universitext, Springer, Berlin.

Kolmogorov, A. N., Petrovskii, I. G., Piskunov, N. S. (1937). Étude de l'equation de la diffusion avec croissance de la quantité de matière et son application à un problème biologique. *Bulletin Université d'Etat à Moscou, Série internationale, section A* 1, 1–25.

Korobeinikov, A., Maini, P. K. (2004). A Lyapunov function and global properties for SIR and SEIR epidemiological models with nonlinear incidence. *Mathematical Biosciences and Engineering* 1(1), 57–60.

Kot, M. (2001). *Elements of mathematical ecology*. Cambridge University Press, Cambridge.

Kot, M., Lewis, M. A., van den Driessche, P. (1996). Dispersal data and the spread of invading organisms. *Ecology* 77(7), 2027–2042.

Kowarik, I. (2003). *Biologische Invasionen: Neophyten und Neozoen in Mitteleuropa*. Ulmer, Stuttgart.

Kribs-Zaleta, C. M. (1999). Core recruitment effects in SIS models with constant total populations. *Mathematical Biosciences* 160, 109–158.

Kribs-Zaleta, C. M. (2004). To switch or taper off: the dynamics of saturation. *Mathematical Biosciences* 192, 137–152.

Kubo, M., Langlais, M. (1991). Periodic solutions for a population dynamics problem with age-dependence and spatial structure. *Journal of Mathematical Biology* 29, 363–378.

Kuznetsov, Y. A. (1995). *Elements of applied bifurcation theory*. Springer-Verlag, Berlin.

Kuznetsov, Y. A., Muratori, S., Rinaldi, S. (1992). Bifurcations and chaos in a periodic predator-prey model. *International Journal of Bifurcation and Chaos* 2, 117–128.

Lambrinos, J. G. (2004). How interactions between ecology and evolution influence contemporary invasion dynamics. *Ecology* 85(8), 2061–2070.

Langford, W. F. (1983). A review of interactions of Hopf and steady-state bifurcations. In *Nonlinear dynamics and turbulence* (Barenblatt, G. I., Iooss, G., Joseph, D. D., eds.). Interaction of Mechanics and Mathematics Series, Pitman, Boston, pp. 215–237.

Langlais, M., Suppo, C. (2000). A remark on a generic SEIRS model and application to cat retroviruses and fox rabies. *Mathematical and Computer Modelling* 31, 117–124.

Lassuy, D. R. (1995). Introduced species as a factor in extinction and endangerment of native fish species. In *Uses and effects of cultured fishes in aquatic ecosystems* (Schramm Jr., H. L., Piper, R. G., eds.). No. 15 in American Fisheries Society Symposium, American Fisheries Society, pp. 391–396.

Lee, K. A., Klasing, K. C. (2004). A role for immunology in invasion biology. *Trends in Ecology & Evolution* 19(10), 523–529.

Leung, B., Drake, J. M., Lodge, D. M. (2004). Predicting invasions: propagule pressure and the gravity of Allee effects. *Ecology* 85(6), 1651–1660.

Leung, B., Lodge, D. M., Finnoff, D., Shogren, J. F., Lewis, M. A., Lamberti, G. (2002). An ounce of prevention or a pound of cure: bioeconomic risk analysis of invasive species. *Proceedings of the Royal Society of London B* 269(1508), 2407–2413.

Levin, S. A., Dushoff, J., Plotkin, J. B. (2004). Evolution and persistence of influenza A and other diseases. *Mathematical Biosciences* 188, 17–28.

Levin, S. A., Segel, L. A. (1976). Hypothesis for origin of planktonic patchiness. *Nature* 259, 659.

Lewis, M. A. (1997). Variability, patchiness, and jump dispersal in the spread of an invading population. In *Spatial ecology. The role of space in population dynamics and interspecific interactions* (Tilman, D., Kareiva, P., eds.). Princeton University Press, Princeton, pp. 46–69.

Lewis, M. A. (2000). Spread rate for a nonlinear stochastic invasion. *Journal of Mathematical Biology* 41, 430–454.

Lewis, M. A., Kareiva, P. (1993). Allee dynamics and the spread of invading organisms. *Theoretical Population Biology* 43, 141–158.

Lewis, M. A., Pacala, S. (2000). Modeling and analysis of stochastic invasion processes. *Journal of Mathematical Biology* 41, 387–429.

Lewis, M. A., van den Driessche, P. (1993). Waves of extinction from sterile insect release. *Mathematical Biosciences* 116, 221–247.

Li, B., Weinberger, H. F., Lewis, M. A. (submitted). Existence of traveling waves for discrete and continuous time cooperative systems.

Li, G., Zhen, J. (2005). Global stability of an SEI epidemic model with general contact rate. *Chaos, Solitons and Fractals* 23, 997–1004.

Li, M. Y., Muldowney, J. S. (1995). Global stability for the SEIR model in epidemiology. *Mathematical Biosciences* 125, 155–164.

Li, M. Y., Wang, L. (2002). Global stability in some SEIR epidemic models. In *Mathematical approaches for emerging and re-emerging infectious diseases. Models, methods, and theory* (Castillo-Chavez, C., Blower, S., van den Driessche, P., Kirschner, D., Yakubu, A.-A., eds.). Springer-Verlag, New York, pp. 295–311.

Liebhold, A., Bascompte, J. (2003). The Allee effect, stochastic dynamics and the eradication of alien species. *Ecology Letters* 6, 133–140.

Liebhold, A. M., Halverson, J. A., Elmes, G. A. (1992). Gypsy moth invasion in North America: a quantitative analysis. *Journal of Biogeography* 19, 513–520.

Lin, J., Andreasen, V., Levin, S. A. (1999). Dynamics of influenza A drift: the linear three-strain model. *Mathematical Biosciences* 162, 33–51.

Liu, W., Hethcote, H. W., Levin, S. A. (1987). Dynamical behavior of epidemiological models with non-linear incidence rate. *Journal of Mathematical Biology* 25, 359–380.

Liu, W., Levin, S. A., Iwasa, Y. (1986). Influence of non-linear incidence rates upon the behaviour of SIRS epidemiological models. *Journal of Mathematical Biology* 23, 187–204.

London, W. P., Yorke, J. A. (1973). Recurrent outbreaks of measles, chickenpox and mumps. I. Seasonal variation in contact rates. *American Journal of Epidemiology* 98, 453–468.

Lotka, A. J. (1925). *Elements of physical biology*. Williams and Wilkins, Baltimore.

Lubina, J. A., Levin, S. A. (1988). The spread of a reinvading species: range expansion in the California sea otter. *The American Naturalist* 131(4), 526–543.

Luther, R. (1906). Räumliche Ausbreitung chemischer Reaktionen. *Zeitschrift für Elektrochemie* 12, 596–600.

Mack, R. N., Simberloff, D., Lonsdale, W. M., Evans, H., Clout, M., Bazzaz, F. A. (2000). Biotic invasions: causes, epidemiology, global consequences, and control. *Ecological Applications* 10(3), 689–710.

Malchow, H. (1993). Spatio-temporal pattern formation in nonlinear nonequilibrium plankton dynamics. *Proceedings of the Royal Society of London B* 251, 103–109.

Malchow, H. (1994). Nonequilibrium structures in plankton dynamics. *Ecological Modelling* 75/76, 123–134.

Malchow, H. (1995). Flow- and locomotion-induced pattern formation in nonlinear population dynamics. *Ecological Modelling* 82, 257–264.

Malchow, H. (1996). Nonlinear plankton dynamics and pattern formation in an ecohydrodynamic model system. *Journal of Marine Systems* 7(2-4), 193–202.

Malchow, H. (1998). Flux-induced instabilities in ionic and population-dynamical interaction systems. *Zeitschrift für Physikalische Chemie* 204, 95–107.

Malchow, H. (2000a). Motional instabilities in predator-prey systems. *Journal of Theoretical Biology* 204, 639–647.

Malchow, H. (2000b). Nonequilibrium spatio-temporal patterns in models of nonlinear plankton dynamics. *Freshwater Biology* 45, 239–251.

Malchow, H., Hilker, F. M., Petrovskii, S. V. (2004a). Noise and productivity dependence of spatiotemporal pattern formation in a prey-predator system. *Discrete and Continuous Dynamical Systems B* 4(3), 705–711.

Malchow, H., Hilker, F. M., Petrovskii, S. V., Brauer, K. (2004b). Oscillations and waves in a virally infected plankton system: Part I. The lysogenic stage. *Ecological Complexity* 1(3), 211–223.

Malchow, H., Hilker, F. M., Sarkar, R. R., Brauer, K. (2005). Spatiotemporal patterns in an excitable plankton system with lysogenic viral infection. *Mathematical and Computer Modelling* (in press).

Malchow, H., Medvinsky, A. B., Petrovskii, S. V. (2004c). Patterns in models of plankton dynamics in a heterogeneous environment. In *Handbook of scaling methods in aquatic ecology: measurement, analysis, simulation* (Seuront, L., Strutton, P. G., eds.). CRC Press, Boca Raton, pp. 401–410.

Malchow, H., Petrovskii, S. V. (2002). Dynamical stabilization of an unstable equilibrium in chemical and biological systems. *Mathematical and Computer Modelling* 36, 307–319.

Malchow, H., Petrovskii, S. V., Hilker, F. M. (2003). Model of spatiotemporal pattern formation in plankton dynamics. *Nova Acta Leopoldina NF* 88(332), 325–340.

Malchow, H., Petrovskii, S. V., Medvinsky, A. B. (2001). Pattern formation in models of plankton dynamics. A synthesis. *Oceanologica Acta* 24(5), 479–487.

Malchow, H., Petrovskii, S. V., Medvinsky, A. B. (2002). Numerical study of plankton-fish dynamics in a spatially structured and noisy environment. *Ecological Modelling* 149, 247–255.

Malchow, H., Radtke, B., Kallache, M., Medvinsky, A. B., Tikhonov, D. A., Petrovskii, S. V. (2000). Spatio-temporal pattern formation in coupled models of plankton dynamics and fish school motion. *Nonlinear Analysis: Real World Applications* 1, 53–67.

Malchow, H., Schimansky-Geier, L. (1985). *Noise and diffusion in bistable nonequilibrium systems.* No. 5 in Teubner-Texte zur Physik, Teubner-Verlag, Leipzig.

Malchow, H., Shigesada, N. (1994). Nonequilibrium plankton community structures in an ecohydrodynamic model system. *Nonlinear Processes in Geophysics* 1(1), 3–11.

Martcheva, M., Castillo-Chavez, C. (2003). Diseases with chronic stage in a population with varying size. *Mathematical Biosciences* 182, 1–25.

Martina, B. E. E., Haagmans, B. L., Kuiken, T., Fouchier, R. A. M., Rimmelzwaan, G. F., van Amerongen, G., Peiris, J. S. M., Lim, W., Osterhaus, A. D. M. E. (2003). Virology: SARS virus infection of cats and ferrets. *Nature* 425, 915.

Maruyama, G. (1955). Continuous Markov processes and stochastic equations. *Rendiconti del Circolo Matematico di Palermo, Series II* 4, 48–90.

May, R. M., Anderson, R. M. (1978). Regulation and stability of host-parasite population interactions. II. Destabilizing processes. *Journal of Animal Ecology* 47, 249–267.

May, R. M., Anderson, R. M. (1979). Population biology of infectious diseases: Part II. *Nature* 280, 455–461.

May, R. M., Lloyd, A. L. (2001). Infection dynamics on scale-free networks. *Physical Review E* 64, 066112.

McCallum, H. (1996). Immunocontraception for wildlife population control. *Trends in Ecology & Evolution* 11(12), 491–493.

McCallum, H. (2000). Achievement and challenge. *Trends in Ecology & Evolution* 15, 352–353.

McCallum, H., Barlow, N., Hone, J. (2001). How should pathogen transmission be modelled? *Trends in Ecology & Evolution* 16(6), 295–300.

McDaniel, L., Houchin, L. A., Williamson, S. J., Paul, J. H. (2002). Lysogeny in *Synechococcus*. *Nature* 415, 496.

McKean, H. P. (1970). Nagumo's equation. *Advances in Mathematics* 4, 209–223.

Meade, D. B. (1992). Qualitative analysis of an epidemic model with directed dispersion. *IMA Preprint Series* 916.

Medvinsky, A. B., Petrovskii, S. V., Tikhonova, I. A., Malchow, H., Li, B.-L. (2002). Spatiotemporal complexity of plankton and fish dynamics. *SIAM Review* 44(3), 311–370.

Medvinsky, A. B., Petrovskii, S. V., Tikhonova, I. A., Venturino, E., Malchow, H. (2001a). Chaos and regular dynamics in a model multi-habitat plankton-fish community. *Journal of Biosciences* 26(1), 109–120.

Medvinsky, A. B., Tikhonov, D. A., Enderlein, J., Malchow, H. (2000). Fish and plankton interplay determines both plankton spatio-temporal pattern formation and fish school walks. A theoretical study. *Nonlinear Dynamics, Psychology and Life Sciences* 4(2), 135–152.

Medvinsky, A. B., Tikhonova, I. A., Aliev, R. R., Li, B.-L., Lin, Z.-S., Malchow, H. (2001b). Patchy environment as a factor of complex plankton dynamics. *Physical Review E* 64(2), 021915:1–7.

Medvinsky, A. B., Tikhonova, I. A., Li, B. L., Malchow, H. (2004). Time delay as a key factor of model plankton dynamics. *Comptes Rendus Biologies* 327(3), 277–282.

Mena-Lorca, J., Hethcote, H. W. (1992). Dynamic models of infectious diseases as regulator of population sizes. *Journal of Mathematical Biology* 30, 693–716.

Menzinger, M., Rovinsky, A. B. (1995). The differential flow instabilities. In *Chemical waves and patterns* (Kapral, R., Showalter, K., eds.). No. 10 in Understanding Chemical Reactivity, Kluwer, Dordrecht, pp. 365–397.

Michaelis, L., Menten, M. L. (1913). Die Kinetik der Invertinwirkung. *Biochemische Zeitschrift* 49, 333–369.

Milner, F. A., Pugliese, A. (1999). Periodic solutions: a robust numerical method for an S-I-R model of epidemics. *Journal of Mathematical Biology* 39, 471–492.

Mitchell, C. E., Power, A. G. (2003). Release of invasive plants from fungal and viral pathogens. *Nature* 421, 625–627.

Moghadas, S. M. (2004). Analysis of an epidemic model with bistable equilibria using the Poincaré index. *Applied Mathematics and Computation* 149, 689–702.

Mollison, D. (1972). The rate of spatial propagation of simple epidemics. In *Proceedings of the 6th Berkeley Symposium on Mathematics, Statistics, and Probability*, vol. 3, pp. 579–614.

Mollison, D. (1977). Spatial contact models for ecological and epidemics spread. *Journal of the Royal Statistical Society B* 39(3), 283–326.

Mollison, D. (1991). Dependence of epidemics and populations velocities on basic parameters. *Mathematical Biosciences* 107, 255–287.

Monod, J., Jacob, F. (1961). General conclusions: Teleonomic mechanisms in cellular metabolism, growth and differentiation. *Cold Spring Harbor Symposia on Quantitative Biology* 26, 389–401.

Moore, C., Newman, M. E. J. (2000). Epidemics and percolation in small-world networks. *Physical Review E* 66, 016128.

Mukherjee, D. (1998). Uniform persistence in a generalized prey-predator system with parasitic infection. *BioSystems* 47, 149–155.

Mukherjee, D. (2003). Persistence in a predator-prey system with diseases in the prey. *Journal of Biological Systems* 11, 101–112.

Murdoch, W. W., Bence, J. (1987). General predators and unstable prey populations. In *Predation: Direct and indirect impacts on aquatic communities* (Kerfoot, W. C., Sih, A., eds.). University Press of New England, Hanover, pp. 17–30.

Murray, J. D. (2002). *Mathematical biology. I: An introduction*. Springer-Verlag, Berlin, 3rd edition.

Murray, J. D. (2003). *Mathematical biology. II: Spatial models and biomedical applications*. Springer-Verlag, Berlin, 3rd edition.

Murray, J. D., Stanley, E. A., Brown, D. L. (1986). On the spatial spread of rabies among foxes. *Proceedings of the Royal Society of London B* 229, 111–150.

Nagasaki, K., Yamaguchi, M. (1997). Isolation of a virus infectious to the harmful bloom causing microalga *Heterosigma akashiwo* (Raphidophyceae). *Aquatic Microbial Ecology* 13, 135–140.

Neal, P. (2003). SIR epidemics on a Bernoulli random graph. *Journal of Applied Probability* 40, 779–782.

Nicolis, G. (1995). *Introduction to nonlinear science*. Cambridge University Press, Cambridge.

Nitzan, A., Ortoleva, P., Ross, J. (1974). Nucleation in systems with multiple stationary states. *Faraday Symposia of The Chemical Society* 9, 241–253.

Noble, J. V. (1974). Geographic and temporal development of plagues. *Nature* 250, 726–729.

Nold, A. (1980). Heterogeneity in disease-transmission modeling. *Mathematical Biosciences* 52, 227–2402.

Okubo, A. (1971). Oceanic diffusion diagrams. *Deep-Sea Research* 18, 789–802.

Okubo, A. (1980). *Diffusion and ecological problems: Mathematical models.* Springer-Verlag, Berlin.

Okubo, A., Levin, S. A. (2001). *Diffusion and ecological problems: Modern perspectives.* Springer-Verlag, New York, 2nd edition.

Okubo, A., Maini, P. K., Williamson, M. H., Murray, J. D. (1989). On the spatial spread of the gray squirrel in Britain. *Proceedings of the Royal Society of London B* 238, 113–125.

O'Neil, L. L., Burkhard, M. J., Hoover, E. A. (1996). Frequent perinatal transmission of Feline Immunodeficiency Virus by chronically infected cats. *Journal of Virology* 70, 2894–2901.

Ortmann, A. C., Lawrence, J. E., Suttle, C. A. (2002). Lysogeny and lytic viral production during a bloom of the cyanobacterium *Synechococcus* spp. *Microbial Ecology* 43, 225–231.

Owen, M. R., Lewis, M. A. (2001). How predation can slow, stop or reverse a prey invasion. *Bulletin of Mathematical Biology* 63, 655–684.

Park, A. W., Gubbins, S., Gilligan, C. A. (2002). Extinction times for closed epidemics: the effects of host spatial structure. *Ecology Letters* 5, 747–755.

Pascual, M. (1993). Diffusion-induced chaos in a spatial predator-prey system. *Proceedings of the Royal Society of London B* 251, 1–7.

Pastor-Satorras, R., Vespignani, A. (2001). Epidemic spreading in scale-free networks. *Physical Review Letters* 86, 3200–3203.

Peaceman, D. W., Rachford Jr., H. H. (1955). The numerical solution of parabolic and elliptic differential equations. *Journal of the Society for Industrial and Applied Mathematics* 3, 28–41.

Peduzzi, P., Weinbauer, M. G. (1993). The submicron size fraction of sea water containing high numbers of virus particles as bioactive agent in unicellular plankton community successions. *Journal of Plankton Research* 15, 1375–1386.

Pertsov, A. M., Davidenko, J. M., Salomonsz, R., Baxter, W., Jalife, J. (1993). Spiral waves of excitation underlie reentrant activity in isolated cardiac muscle. *Circulation Research* 72, 631–650.

Petrovskii, S., Li, B.-L. (2003). An exactly solvable model of population dynamics with density-dependent migrations and the Allee effect. *Mathematical Biosciences* 186, 79–91.

Petrovskii, S., Morozov, A., Li, B.-L. (2005a). Regimes of biological invasion in a predator-prey system with the Allee effect. *Bulletin of Mathematical Biology* 67, 637–661.

Petrovskii, S., Shigesada, N. (2001). Some exact solutions of a generalized Fisher equation related to the problem of biological invasion. *Mathematical Biosciences* 172, 73–94.

Petrovskii, S. V., Malchow, H. (1999). A minimal model of pattern formation in a prey-predator system. *Mathematical and Computer Modelling* 29, 49–63.

Petrovskii, S. V., Malchow, H. (2000). Critical phenomena in plankton communities: KISS model revisited. *Nonlinear Analysis: Real World Applications* 1, 37–51.

Petrovskii, S. V., Malchow, H. (2001). Wave of chaos: new mechanism of pattern formation in spatio-temporal population dynamics. *Theoretical Population Biology* 59(2), 157–174.

Petrovskii, S. V., Malchow, H., Hilker, F. M., Venturino, E. (2005b). Patterns of patchy spread in deterministic and stochastic models of biological invasion and biological control. *Biological Invasions* 7, 771–793.

Petrovskii, S. V., Malchow, H., Li, B.-L. (2005c). An exact solution of a difffusive predator-prey system. *Proceedings of the Royal Society of London A* 461, 1029–1053.

Petrovskii, S. V., Morozov, A. Y., Venturino, E. (2002a). Allee effect makes possible patchy invasion in a predator-prey system. *Ecology Letters* 5, 345–352.

Petrovskii, S. V., Vinogradov, M. E., Morozov, A. Y. (2002b). Formation of the patchiness in the plankton horizontal distribution due to biological invasion in a two-species model with account for the Allee effect. *Oceanology* 42(3), 363–372.

Pimentel, D. (1986). Biological invasions of plants and animals in agriculture and forestry. In *Ecology of biological invasions of North America and Hawaii* (Mooney, H. A., Drake, J. A., eds.). Springer-Verlag, New York, pp. 149–162.

Pimentel, D. (ed.) (2002). *Biological invasions: Economic and environmental costs of alien plant, animal, and microbe species*. CRC Press, Boca Raton.

Pimentel, D., Lach, L., Zuniga, R., Morrison, D. (2000). Environmental and economic costs of non-indigenous species in the United States. *BioSciences* 50, 53–65.

Pontier, D., Auger, P., Bravo de la Parra, R., Sánchez, E. (2000). The impact of behavioral plasticity at individual level on domestic cat population dynamics. *Ecological Modelling* 133, 117–124.

Prenter, J., MacNeil, C., Dick, J. T. A., Dunn, A. M. (2004). Roles of parasites in animal invasions. *Trends in Ecology & Evolution* 19(7), 385–390.

Pugliese, A. (1990). Population models for diseases with no recovery. *Journal of Mathematical Biology* 28, 65–82.

Pugliese, A. (1991). An S→E→I epidemic model with varying total population size. In *Differential equation models in biology, epidemiology and ecology* (Busenberg, S. N., Martelli, M., eds.). No. 92 in Lecture Notes in Biomathematics, Springer-Verlag, Berlin, pp. 121–138.

Rass, L., Radcliff, J. (2003). *Spatial deterministic epidemics*. American Mathematical Society, Providence RI.

Reiser, W. (1993). Viruses and virus like particles of freshwater and marine eukaryotic algae – a review. *Archiv für Protistenkunde* 143, 257–265.

Rinaldi, S., Muratori, S., Kuznetsov, Y. (1993). Multiple attractors, catastrophes and chaos in seasonally perturbed predator-prey communities. *Bulletin of Mathematical Biology* 55, 15–35.

Rogers, A. B., Hoover, E. A. (1998). Maternal-fetal Feline Immunodeficiency Virus transmission: timing and tissue tropisms. *The Journal of Infectious Diseases* 178, 960–967.

Ross, R. (1911). *The prevention of malaria*. Murray, London, 2nd edition.

Roughgarden, J. (1975). Evolution of marine symbiosis – a simple cost-benefit model. *Ecology* 56, 1201–1208.

Rovinsky, A. B., Menzinger, M. (1992). Chemical instability induced by a differential flow. *Physical Review Letters* 69, 1193–1196.

Ruan, S., Wang, W. (2003). Dynamical behaviour of an epidemic model with a nonlinear incidence rate. *Journal of Differential Equations* 188, 135–163.

Sala, O. E., Chapin, F. S., Armesto, J. J., Berlow, E., Bloomfield, J., Dirzo, R., Huber-Sanwald, E., Huenneke, L. F., Jackson, R. B., King, A., Leemans, R., Lodgeand, D. M., Mooney, H. A., Oesterheld, M., Poff, N. L., Sykes, M. T., Walker, B. H., Walker, M., Wall, D. H. (2000). Global biodiversity scenarios for the year 2100. *Science* 287, 1770–1774.

Sarkar, R. R., Chattopadhayay, J. (2003). Occurence of planktonic blooms under environmental fluctuations and its possible control mechanism – mathematical models and experimental observations. *Journal of Theoretical Biology* 224, 501–516.

Satnoianu, R. A., Menzinger, M. (2000). Non-Turing stationary patterns in flow-distributed oscillators with general diffusion and flow rates. *Physical Review E* 62(1), 113–119.

Satnoianu, R. A., Menzinger, M., Maini, P. K. (2000). Turing instabilities in general systems. *Journal of Mathematical Biology* 41, 493–512.

Sato, K., Matsuda, H., Sasaki, A. (1994). Pathogen invasion and host extinction in lattice structured populations. *Journal of Mathematical Biology* 32, 251–268.

Scheffer, M. (1991a). Fish and nutrients interplay determines algal biomass: a minimal model. *Oikos* 62, 271–282.

Scheffer, M. (1991b). Should we expect strange attractors behind plankton dynamics - and if so, should we bother? *Journal of Plankton Research* 13, 1291–1305.

Scheffer, M. (1998). *Ecology of shallow lakes*. Chapman & Hall, London.

Scheffer, M., Rinaldi, S., Kuznetsov, Y. A., van Nes, E. H. (1997). Seasonal dynamics of daphnia and algae explained as a periodically forced predator-prey system. *Oikos* 80, 519–532.

Schenzle, D. (1984). An age structured model of pre- and post-vaccination measles transmission. *IMA Journal of Mathematics Applied in Medicine and Biology* 1, 169–191.

Scherm, H. (1996). On the velocity of epidemic waves in model plant disease epidemics. *Ecological Modelling* 87, 217–222.

Segel, L. A., Jackson, J. L. (1972). Dissipative structure: an explanation and an ecological example. *Journal of Theoretical Biology* 37, 545–559.

Sellon, R. K., Jordan, H. L., Kennedy-Stoskopf, S., Tompkins, M. B., Tompkins, W. A. (1994). Feline Immunodeficiency Virus can be experimentally transmitted via milk during acute maternal infection. *Journal of Virology* 68, 3380–3385.

Sherratt, J. A., Eagan, B. T., Lewis, M. A. (1997). Oscillations and chaos behind predator-prey invasion: mathematical artifact or ecological reality? *Philosophical Transactions of the Royal Society of London B* 352, 21–38.

Sherratt, J. A., Lewis, M. A., Fowler, A. C. (1995). Ecological chaos in the wake of invasion. *Proceedings of the National Academy of Sciences of the United States of America* 92, 2524–2528.

Shigesada, N., Kawasaki, K. (1997). *Biological invasions: Theory and practice*. Oxford University Press, Oxford.

Shigesada, N., Kawasaki, K. (2002). Invasion and the range expansion of species: effects of long-distance dispersal. In *Dispersal ecology* (Bullock, J., Kenward, R., Hails, R., eds.). Blackwell Science, Malden MA, pp. 350–373.

Shigesada, N., Kawasaki, K., Takeda, Y. (1995). Modeling stratified diffusion in biological invasions. *The American Naturalist* 146(2), 229–251.

Shigesada, N., Kawasaki, K., Teramoto, E. (1986). Traveling periodic waves in heterogeneous environments. *Theoretical Population Biology* 30, 143.

Simberloff, D., Gibbons, L. (2004). Now you see them, now you don't! - Population crashes of established introduced species. *Biological Invasions* 6, 161–172.

Simberloff, D., Parker, I. M., Windle, P. N. (2005). Introduced species policy, management, and future research needs. *Frontiers in Ecology and the Environment* 3, 12–20.

Simberloff, D., Stiling, P. (1996a). How risky is biological control? *Ecology* 77, 1965–1974.

Simberloff, D., Stiling, P. (1996b). Risks of species introduced for biological control. *Biological Conservation* 78, 185–192.

Singh, B. K., Chattopadhyay, J., Sinha, S. (2004). The role of virus infection in a simple phytoplankton zooplankton system. *Journal of Theoretical Biology* 231, 153–166.

Skellam, J. G. (1951). Random dispersal in theoretical populations. *Biometrika* 38, 196–218.

Smith, H. L. (1986). Cooperative systems of differential equations with concave nonlinearities. *Nonlinear Analysis* 10, 1037–1052.

Snow, A. A., Andow, D. A., Gepts, P., Hallerman, E. M., Power, A., Tiedje, J. M., Wolfenbarger, L. L. (2005). Genetically engineered organisms and the environment: current status and recommendations. *Ecological Applications* 15(2), 377–404.

Soper, H. E. (1929). Interpretation of periodicity in disease prevalence. *Journal of the Royal Statistical Society B* 92, 34–73.

Sparger, E. E. (1993). Current thoughts of Feline Immunodeficiency Virus infection. *Veterinary Clinics of North America: Small Animal Practice* 23, 173–191.

Stech, H. W., Williams, M. (1981). Stability for a class of cyclic epidemic models with delay. *Journal of Mathematical Biology* 11, 95–103.

Steele, J., Henderson, E. W. (1981). A simple plankton model. *The American Naturalist* 117, 676–691.

Steele, J., Henderson, E. W. (1992a). A simple model for plankton patchiness. *Journal of Plankton Research* 14, 1397–1403.

Steele, J. H., Henderson, E. W. (1992b). The role of predation in plankton models. *Journal of Plankton Research* 14, 157–172.

Steffen, E., Malchow, H., Medvinsky, A. B. (1997). Effects of seasonal perturbation on a model plankton community. *Environmental Modeling and Assessment* 2, 43–48.

Stephens, P. A., Sutherland, W. J. (1999). Consequences of the Allee effect for behaviour, ecology and conservation. *Trends in Ecology & Evolution* 14(10), 401–405.

Stephens, P. A., Sutherland, W. J., Freckleton, R. P. (1999). What is the Allee effect? *Oikos* 87(1), 185–190.

Suppo, C., Naulin, J.-M., Langlais, M., Artois, M. (2000). A modelling approach to vaccination and contraception programmes for rabies control in fox populations. *Proceedings of the Royal Society of London B* 267, 1575–1582.

Susser, E. (2004). Eco-epidemiology: Thinking outside the black box. *Epidemiology* 15(5), 519–520.

Susser, M., Susser, E. (1996). Choosing a future for epidemiology: II. From black box to Chinese boxes and eco-epidemiology. *American Journal of Public Health* 86(5), 674–677.

Suttle, C. A. (2000). Ecological, evolutionary, and geochemical consequences of viral infection of cyanobacteria and eukaryotic algae. In *Viral ecology* (Hurst, C. J., ed.). Academic Press, San Diego, pp. 247–296.

Suttle, C. A., Chan, A. M. (1993). Marine cyanophages infecting oceanic and coastal strains of *Synechococcus*: abundance, morphology, cross-infectivity and growth characteristics. *Marine Ecology Progress Series* 92, 99–109.

Suttle, C. A., Chan, A. M., Cottrell, M. T. (1990). Infection of phytoplankton by viruses and reduction of primary productivity. *Nature* 347, 467–469.

Suttle, C. A., Chan, A. M., Feng, C., Garza, D. R. (1993). Cyanophages and sunlight: a paradox. In *Trends in microbial ecology* (Guerrero, R., Pedros-Alio, C., eds.). Spanish Society for Microbiology, Barcelona, pp. 303–307.

Takasu, F., Yamamoto, N., Kawasaki, K., Togashi, K., Kishi, Y., Shigesada, N. (2000). Modeling the expansion of an introduced tree disease. *Biological Invasions* 2, 141–150.

Tarutani, K., Nagasaki, K., Yamaguchi, M. (2000). Viral impacts on total abundance and clonal composition of the harmful bloom-forming phytoplankton *Heterosigma akashiwo*. *Applied and Environmental Microbiology* 66(11), 4916–4920.

Thieme, H. R. (1992). Epidemic and demographic interaction in the spread of potentially fatal diseases in growing populations. *Mathematical Biosciences* 111, 99–130.

Thieme, H. R. (2003). *Mathematics in population biology*. Princeton University Press, Princeton NJ.

Thieme, H. R., Castillo-Chavez, C. (1993a). How may infection-age-dependent infectivity affect the dynamics of HIV/AIDS? *SIAM Journal on Applied Mathematics* 53(5), 1447–1479.

Thieme, H. R., Castillo-Chavez, C. (1993b). On the role of variable infectivity in the dynamics of HIV epidemic. In *Mathematical and statistical approaches to AIDS epidemiology* (Castillo-Chavez, C., ed.). No. 83 in Lecture Notes in Biomathematics, Springer, Berlin, pp. 157–176.

Thomas, J. W. (1995). *Numerical partial differential equations: Finite difference methods*. Springer-Verlag, New York.

Tompkins, D. M., White, A. R., Boots, M. (2003). Ecological replacement of native red squirrels by invasive greys driven by disease. *Ecology Letters* 6, 189–196.

Torchin, M. E., Lafferty, K. D., Dobson, A. P., McKenzle, V. J., Kuris, A. M. (2003). Introduced species and their missing parasites. *Nature* 421, 628–630.

Torchin, M. E., Lafferty, K. D., Kuris, A. M. (2002). Parasites and marine invasions. *Parasitology* 124, S137–S151.

Truscott, J. E., Brindley, J. (1994a). Equilibria, stability and excitability in a general class of plankton population models. *Philosophical Transactions of the Royal Society of London A* 347, 703–718.

Truscott, J. E., Brindley, J. (1994b). Ocean plankton populations as excitable media. *Bulletin of Mathematical Biology* 56, 981–998.

Turchin, P. (2003). *Complex population dynamics. A theoretical/empirical synthesis*. Princeton University Press, Princeton NJ.

Turchin, P., Ellner, S. P. (2000). Living on the edge of chaos: population dynamics of Fennoscandian volves. *Ecology* 81(11), 3099–3116.

Turing, A. M. (1952). On the chemical basis of morphogenesis. *Philosophical Transactions of the Royal Society of London B* 237, 37–72.

Ueland, K., Nesse, L. L. (1992). No evidence of vertical transmission of naturally acquired Feline Immunodeficiency Virus infection. *Veterinary Immunology and Immunopathology* 33, 301–308.

van den Bosch, F., Metz, J. A. J., Diekmann, O. (1990). The velocity of spatial population expansion. *Journal of Mathematical Biology* 28, 529–565.

van den Driessche, P., Watmough, J. (2000). A simple SIS model with a backward bifurcation. *Journal of Mathematical Biology* 40, 525–540.

van den Driessche, P., Watmough, J. (2003). Epidemic solutions and endemic catastrophies. In *Dynamical systems and their applications in biology* (Ruan, S., Wolkowicz, G. S. K., Wu, J., eds.). No. 36 in Fields Institute Communications, American Mathematical Society, Providence, Rhode Island, pp. 247–257.

van Etten, J. L., Lane, L. C., Meints, R. H. (1991). Viruses and viruslike particles of eukaryotic algae. *Microbiological Reviews* 55, 586–620.

van Saarlos, W. (2003). Front propagation into unstable states. *Physics Reports* 386, 29–222.

Vander Zanden, M. J. (2005). The success of animal invaders. *Proceedings of the National Academy of Sciences of the United States of America* 102(20), 7055–7056.

Veit, R. R., Lewis, M. A. (1996). Dispersal, population growth, and the Allee effect: Dynamics of the House Finch invasion of eastern North America. *The American Naturalist* 148(2), 255–274.

Venturino, E. (1995). Epidemics in predator-prey models: disease in the prey. In *Mathematical population dynamics: Analysis of heterogeneity. Volume one: Theory of epidemics* (Arino, O., Axelrod, D., Kimmel, M., Langlais, M., eds.). Wuerz, Winnipeg, pp. 381–393.

Venturino, E. (2001). The effect of diseases on competing species. *Mathematical Biosciences* 174, 111–131.

Vinogradov, M. E., Shushkina, E. A., Anochina, L. L., Vostokov, S. V., Kucheruk, N. V., Lukashova, T. A. (2000). Mass development of ctenophore Beroe ovata Eschscholtz off the north-eastern coast of the Black Seas. *Oceanology* 40, 52–55.

Volterra, V. (1926). Variazioni e fluttuazioni del numero d'individui in specie animali conviventi. *Memorie dell'Accademia dei Lincei* III(6), 31–113.

Wang, L., Li, J. (2005). Global stability of an epidemic model with nonlinear incidence rate and differential infectivity. *Applied Mathematics and Computation* 161, 769–778.

Wang, M.-H., Kot, M. (2001). Speeds of invasion in a model with strong or weak Allee effects. *Mathematical Biosciences* 171(1), 83–97.

Wang, M.-H., Kot, M., Neubert, M. G. (2002). Integrodifference equations, Allee effects, and invasions. *Journal of Mathematical Biology* 44, 150–168.

Weinberger, H. F. (1978). Asymptotic behavior of a model in population genetics. In *Nonlinear partial differential equations and applications* (Chadam, J. M., ed.). No. 648 in Lecture Notes in Mathematics, Springer-Verlag, Berlin, pp. 47–96.

Weinberger, H. F. (1982). Long-time behavior of a class of biological models. *SIAM Journal of Mathematical Analysis* 13, 353–396.

Wiggins, S. (2003). *Introduction to applied nonlinear dynamical systems and chaos*. Springer-Verlag, New York, 2nd edition.

Wilcox, R., Fuhrman, J. (1994). Bacterial viruses in coastal seawater: lytic rather than lysogenic production. *Marine Ecology Progress Series* 114, 35–45.

Wilhelm, S. W., Suttle, C. A. (1999). Viruses and nutrient cycles in the sea. *BioScience* 49(10), 781–788.

Williamson, M. (1996). *Biological invasions*. Chapman & Hall, London.

Wilson, E. B., Worcester, J. (1945a). The law of mass action in epidemiology. *Proceedings of the National Academy of Sciences of the United States of America* 31, 24–34.

Wilson, E. B., Worcester, J. (1945b). The law of mass action in epidemiology, II. *Proceedings of the National Academy of Sciences of the United States of America* 31, 109–116.

Wodarz, D., Sasaki, A. (2004). Apparent competition and recovery from infection. *Journal of Theoretical Biology* 227, 403–412.

Wommack, K. E., Colwell, R. R. (2000). Virioplankton: viruses in aquatic ecosystems. *Microbiology and Molecular Biology Reviews* 64(1), 69–114.

Xiao, Y., Chen, L. (2001a). Analysis of a three-species eco-epidemiological model. *Journal of Mathematical Analysis and Applications* 258, 733–754.

Xiao, Y., Chen, L. (2001b). Modeling and analysis of a predator-prey model with disease in the prey. *Mathematical Biosciences* 171, 59–82.

Yachi, S., Kawasaki, K., Shigesada, N., Teramoto, E. (1989). Spatial patterns of propagating waves of fox rabies. *Forma* 4, 3–12.

Yamamoto, J. K., Hansen, H., Ho, E. W., Morishita, T. Y., Okuda, T., Sawa, T. R., Nakamura, R. M., Pedersen, N. C. (1989). Epidemiological and clinical aspects on Feline Immunodeficiency Virus infection in cats from continental United States and Canada and possible modes of transmission. *Journal of the American Veterinary Medical Association* 194(2), 213–220.

Yamamoto, J. K., Pedersen, N. C., Ho, E. W., Okuda, T., Theilen, G. H. (1988). Feline immunodeficiency syndrome: a comparison between T-lymphotropic lentivirus and Feline Leukemia Virus. *Leukemia* 2, 204–215.

Yamamura, N. (1996). Evolution of mutualistic symbiosis: a differential equation model. *Researches on Population Ecology* 38(2), 211–218.

Zenger, E. (1992). An update on FeLV and FIV: The diagnosis prevention and treatment. *Veterinary Medicine* 3, 202–210.

Zhang, J., Ma, Z. (2003). Global dynamics of an SEIR epidemic model with saturating contact rate. *Mathematical Biosciences* 185, 15–32.

Zhdanov, V. P. (2003). Propagation of infection and the predator-prey interplay. *Journal of Theoretical Biology* 225, 489–492.

Zhou, J., Hethcote, H. W. (1994). Population size dependent incidence in models for diseases without immunity. *Journal of Mathematical Biology* 32, 809–834.